SLOW
LIVING

Vandana Shiva is a world-renowned environmental thinker, activist, feminist, and philosopher of science whose work has shaped global movements for ecological justice. Trained as a physicist at Panjab University, she earned her PhD from the University of Western Ontario before turning to interdisciplinary research in ecology, science policy, and sustainability. She is the founder of the Research Foundation for Science, Technology and Ecology and Navdanya, and serves as President of Navdanya International. Her advocacy for seed freedom, biodiversity, and food sovereignty has influenced policy and grassroots movements worldwide. She is the recipient of major international honours, including the Right Livelihood Award (1993), often referred to as the 'Alternative Nobel Prize,' the Sydney Peace Prize (2010), TIME magazine's 'Environmental Hero' recognition, the UN Global 500 Roll of Honour, the Fukuoka Prize, and the Order of the Golden Ark.

Shreya Jani is a peace educator and cultural practitioner who began her journey at the age of 15, working with Vandana Shiva, an association that has continued for over three decades.

Her work spans geographies, demographies, and contexts, engaging youth and educators in peace-building across Kashmir, Manipur, Afghanistan, Cyprus, and beyond, through social and emotional learning, peace education, and ecological literacy, in association with NGOs, INGOs, and public institutions including NCERT and UNESCO.

She is also the founder of Sunshine Masala Witch, an ethical, nature-rooted business that weaves joy, ritual, and care into everyday life while creating meaningful livelihood and leadership opportunities for young women.

SLOW LIVING

What you can do about
CLIMATE CHANGE

Vandana Shiva
Shreya Jani

ROLI

First published by Roli Books in India in 2026

Roli Books Pvt. Ltd.
M-75, Greater Kailash II Market, New Delhi 110 048
Phone: +91 (011) 40682000
E-mail: info@rolibooks.com
Website: www.rolibooks.com
Also at Chennai & Mumbai

ISBN: 9789349474079

Typeset in Olympian LT Std by Roli Books

CONTENTS

1

THE ART *of* SLOWING DOWN

Discovering the Joy *of* Being

Have you ever noticed how time feels scarce even when our lives are full? How days pass quickly, yet leave us tired rather than fulfilled? How moments of joy rarely arrive with a sense of calm? Many of us move through our lives feeling hurried, stretched, and slightly behind – as if we are always trying to catch up with something just out of reach.

This is not simply a personal failure of planning or discipline. It is a shared condition of our time. We live with a persistent sense that there is never enough time – for rest, for care, for reflection, or for what truly matters. This quiet exhaustion has become so familiar that we rarely question it. Yet it shapes how we work, how we relate to one another, and how we treat the Earth itself.

The era we live in today – let us call it the *Fossil Age* – is organized around speed. Our economies, habits, and expectations are powered by carbon the Earth took more than 600 million years to form. What accumulated slowly over deep geological time is extracted and burned in moments. Living within this compressed tempo, we have come to treat urgency as normal and slowness as a problem to be solved.

This way of living has subtly altered how we experience time. Days feel packed yet thin. Efficiency replaces attentiveness. We measure success by how much can be done, rather than by what can be sustained. Time becomes something to manage, save, or conquer – rather than something to inhabit.

For much of modern history, this experience of time has been reinforced by a particular worldview. In the old Newtonian–Cartesian mechanical framework, time is linear, one-dimensional, and separate from space. It flows forward like a neutral backdrop against which life unfolds. Ancient wisdom traditions – and later Einstein's Theory of Relativity – offer a very different understanding: time and space are inseparable, woven together as a living whole.

This understanding of unity extends beyond physics. It reminds us that humans are not separate from the Earth, nor from one another. Life does not consist of isolated individuals moving through empty time, competing for advantage. It unfolds through relationships – between people, communities, species, and ecosystems – within shared rhythms.

When this sense of oneness is forgotten, separation becomes normal. Nature becomes a resource. Time becomes a commodity. Care is replaced by extraction, and power begins to concentrate in the hands of the few.

Because time and space are intertwined, pressure on one is always felt on the other. When speed becomes the dominant measure of success, processes that require time are denied it. Forests are cleared faster than they can regenerate. Soils are depleted faster than they can recover. Animals are pushed into crowded, artificial environments. What we call efficiency in time often results in a steady loss of space – less room for renewal, resilience, and coexistence.

These spatial compressions have consequences we can no longer ignore. As ecosystems are disrupted and species are forced into closer, more fragile contact with one another, new diseases emerge. Over the past 50 years, more than 300 new pathogens have appeared. Around 70 percent of human diseases – including HIV, Ebola, influenza, MERS, SARS, and the coronavirus of 2020 – originated when viruses crossed from animals to humans following ecological disturbance. These outbreaks are not accidents; they are signals of systems pushed beyond their capacity to absorb stress.

Seen this way, the Fossil Age is not only an energy system – it is a way of organizing life. It equates speed with progress and extraction with efficiency. This logic took a definitive form under industrial capitalism, most visibly through the Empire of Cotton, which reorganized land, labour, and agriculture around distant markets and relentless production.

It was this trajectory that **Mahatma Gandhi** warned against when he said:

> God forbid that India should ever take to industrialization after the manner of the West.... If an entire nation took to similar economic exploitation, it would strip the world bare like locusts.

Gandhi recognized that the danger was not only economic or political, but civilizational. The Cotton Empire belongs to the past, yet its logic persists today – in global supply chains, fast fashion, industrial agriculture, and systems that treat land, labour, and time as disposable.

His response was neither nostalgia nor rejection of work. Gandhi placed the spinning wheel – the *charkha* – at the centre of daily life. Spinning was a way of restoring proportion: between effort and outcome, need and excess, human bodies and the materials they worked with. It slowed production to a pace that could be lived with, cared for, and sustained.

Even today, women who spin khadi describe the practice as meditative – grounded in repetition, patience, and presence. The spinning wheel endures not as a relic of the past, but as a question for the present: in a world driven by speed and disposability, what are the practices that can return rhythm to our lives?

Scientific assessments now confirm what this loss of rhythm produces. Species are disappearing at unprecedented rates. Ecological systems are becoming increasingly fragile. These are not isolated crises, but expressions of a deeper imbalance – systems organized for short-term gain rather than long-term continuity.

This imbalance was articulated with particular clarity in the **International Assessment of Agricultural Knowledge, Science and Technology for Development** (IAASTD) Synthesis Report, the most comprehensive review of the global agricultural system to date. The assessment brought together more than 400 researchers and authors from 61 countries through a multi-stakeholder process guided by a bureau co-sponsored by the **Food and Agriculture Organization**, the **Global Environment Facility**, the **United Nations Development Programme**, the **United Nations Environment Programme**, **UNESCO**, the **World Bank**, and the **World Health Organization**. According to statistics from this document:

• Water withdrawal from the rivers 50 years ago was a third of what it is today.

• Agriculture consumes 75 percent of global freshwater worldwide.

• Increasing pollution also contributes to water quality problems affecting rivers and streams.

• Agriculture contributes about 60 percent of anthropogenic emissions of CH_4 (methane) and about 50 percent of N_2O (nitrous oxide) emissions.

• Inappropriate fertilization has led to eutrophication and large dead zones in several coastal areas.

• Inappropriate use of pesticides has led to groundwater pollution, health issues, and loss of biodiversity.

Taken together, these findings tell a simple story. Life struggles when everything is pushed to move faster than it can recover. Systems built around speed, scale, and short-term output strain the conditions that allow renewal. This is not about blame or failure. It is about noticing that something is out of step.

Slowing down, in this sense, is not about stepping away from the world. It is about coming back into it. At a slower pace, connections begin to reappear and regenerate – between food and soil, work and well-being, time and attention. Questions that once felt separate start to belong together: how we farm, how we eat, what we wear, how we build, and how we spend our days.

Much of what people feel today – tiredness, restlessness, a sense of always being behind – comes from carrying too much for too long without pause. The planet feels this too. When there is no time to regenerate, things thin and fray. Slowing down does not promise perfection. It offers something quieter and more useful: the chance to mend, to realign, to make life workable again.

Change rarely begins with grand gestures. It starts in ordinary places – in the choices we repeat, the pace we set, the care we allow ourselves and others. On their own, these choices may seem small. Over time, they quietly reshape how we live with time, with nature, and with one another.

This book offers ideas, stories, practices, and ways of doing things – but it does not ask you to follow them in order, or all at once. You can begin anywhere, move slowly, skip what does not fit, and return when something calls you back. Let it work the way real change does: unevenly, patiently, in its own time.

What follows is not a set of instructions, but a practice – one that begins wherever you are, with the life already in your hands. Let this be a companion you walk with, not a path you must complete – something to return to, absorb gradually, and live into over time.

Slow Living

'How could we even begin to disarm greed and envy? Perhaps by being much less greedy and envious ourselves; perhaps by resisting the temptation of letting our luxuries become needs; and perhaps by even scrutinizing our needs to see if they cannot be simplified and reduced.'
ERNST F. SCHUMACHER
Small Is Beautiful: A Study of Economics as if People Mattered

According to Native American wisdom, 'things are thieves of time'. This statement has never held truer than today, when the measure of human existence has been reduced to the ability of a person to accumulate and amass, and then discard at an equally alarming rate. Life in this fast lane has created many global challenges, and it is time for humanity to slow down and introspect.

Why are we always in such a rush? What is the cure for this time sickness, and what is this paradox of fast versus slow?

Let's simplify it and review the 10 differences between Fast Living and Slow Living.

FAST LIVING	SLOW LIVING
Quantity Fast food, fast fashion, fast recreation – all of it needs one to be fast consumers with an insatiable hunger for material consumption.	**Quality** Slow consumption makes us reflect on buying less in a holistic way, so that the health of people who produce it, the planet that we live on, and our own are accounted for in consumption decisions.
Monoculture Mass production needs standardization and monoculture of raw material and production to survive.	**Diversity in nature as the norm** No two leaves are the same, and thus, no two handmade products are the same.
Mechanical labour People are a cog in the wheel of large production.	**Bread labour** Pride in what you create, for you are essential to the creation of the product.
Disposable Use and throw is the mantra of a fast life.	**Repairable** Reuse, repurpose, upcycle, and being ecological are the mantras of slow life.
Instant gratification drives the fast world.	**Patience and care** are cultivated by slowing down.
Web of lies Clever marketing and need creation for even the most useless things is the lynchpin of fast living.	**Web of life** Understanding the interconnectivity of life and existence by reflective rather than reactive existence is the cornerstone of slow living.

Keeping up with the Joneses Showing off to neighbours guides life in the fast lane.	**Keeping up with Gaia** Being accountable to the Earth guides the slow path of living.
Greed for more things is ingrained in fast living.	**Need-based** consumption is the guiding light for slow living.
Individualistic and anthropocentric The self and its every whim and fancy are at the centre of fast existence.	**Collective reason and the celebration of interbeing** *'You are, therefore I am'* is embedded in the consciousness of slow existence.
Divisive approach, creating disparities, feel different and unique are essential for fast living.	**Inclusive** approach, care, and compassion for everyone is embedded in the slow life.

The statements above are bold claims, and one may be sceptical about them at a glance. But we hope that as you take this journey with us in the following chapters and get encouraged to walk the talk, in your own time, at your own pace, you will see the lotus of mindfulness and love for the Earth bloom.

Slowing Down & Embracing Simplicity

10 STEPS TOWARDS A 'SIMPLER LIFE'

'To live more simply is to live more purposefully and with a minimum of needless distraction.... To live more simply is to unburden ourselves – to live more lightly, cleanly and aerodynamically. It is to establish a more direct, unpretentious and unencumbered relationship with all aspects of our lives: The things that we consume, the work that we do, our relationships with others, our connections with nature and the cosmos, and more. Simplicity of living means meeting life face-to-face.'

DUANE ELGIN

Voluntary Simplicity: Toward a Way of Life that is Outwardly Simple, Inwardly Rich

1. CONSIDER YOUR REASONS

The purpose of simple living is to declutter your mind and space and understand what is important and what is not. So, before you can start simplifying your life, you need to figure out what is most crucial in your life. There are no universal truths to living simply – it is merely a question of deciding what you want and why, and then moving forward. By reflecting on the ripple effects of the choices we make, our community, and the Earth can help us align our reasons with the kind of world we want to live in.

2. SPEND MINDFULLY

Once you have decided the purpose of your simple life, you can start directing your consumption in ways that match your reasons. Before you buy anything, ask yourself how it fits with your new vision of life. For example, if you are aiming to live in harmony with the earth, then asking this question may lead you not to buy imported oranges from the market, as it adds to your food miles and carbon footprint. On the other hand, if your goal is to reduce your work hours, you might buy fewer shoes and clothes, for you will realize that it is not worth the hours of work it would take you to pay for them. There are no right or wrong reasons. There are only reasons and motivations. Just take a minute to pause to reflect if everything you buy is really worthwhile and adding value to your life.

3. DE-CLUTTER

Cleansing your home of excessive stuff is a good step towards a simple life. It can help you realize how much of what you own is not truly necessary or that it's just not of any use at all. This can clear up space so that you have only the things you really do need and value around you. Start with your immediate surroundings. Pick up objects and ask yourself, 'Do I ever use this? Does it have value for me? Does it fit into my simpler life?' If you can't answer yes to all three questions, put it in the donation box or for recycling.

4. LESS IS MORE

One advantage of consciously consuming is that you end up with fewer belongings. Hence, what you possess are things that you value and cherish. It will make you repair things when they are broken and reduce what you buy, teaching you how to care for the planet and people more.

5. WITNESS INTERCONNECTIVITY

The Hindi word for philosophy is *darshan*, which means 'to see' or 'witness'. The act of slowing down thus invites us to awaken the inner observer and find joy and compassion in witnessing the interconnectivity of the web of life. It helps us understand that the clothes we wear and the food we eat are connected to everything – people, land, water, and other sentient beings. They are connected to systems that either rob or enable them with every choice we make in our daily lives. This understanding of interconnectivity is a key ingredient to slow, compassionate living.

6. RETURNING TO THE CARE OF THE COMMONS

The Commons and communities are beyond the state and the market. Community resources, or the Commons, are not just used collectively by a particular community but in a way that ensures their sustenance for future generations. This sustainability is ensured through a series of practices and knowledge that are indigenous to these groups and are culturally significant for a given area and community – the control and exploitation of which not only leads to depletion of these resources but also puts the community at risk. Thus, looking after the Commons, such as water, soil, air, and the ecosystem, must be an act of daily life. Creating community gardens, working towards water-management systems, and reducing pollution of all kinds are not event-based exercises but must enter our daily actions.

7. CONSCIOUS EATING

It means eating simple, home-cooked meals made with unprocessed, local, and organic foods. Making a conscious choice to spend more on local and Fair Trade products, as they benefit health, people, and the planet. You can opt to grow your own vegetables and fruits, investing time instead of money in what you eat.

8. WORK-LIFE BALANCE

Slowing down and living simply isn't just about spending less money. It helps you live the kind of life you want. You may work longer hours to earn more money, but if you have fewer expenses in the first place, you can choose to do work which is less demanding on your time, leaving you with more to do what you truly want to. Rather than aiming to increase your standard of living with a higher salary, you can focus on improving your quality of life.

9. APPRECIATION AND GRATITUDE

Having less makes one more appreciative of one's belongings. These little moments of appreciation for what you have and truly enjoy make the simple life a more satisfying one.

10. LEARN MORE

There are a number of online resources that can provide you with information about the simple life and how to achieve it. Often, words like 'downshifting' or 'ecological living' are used to describe making this shift. Reading and watching other people's experiences can inspire one to take the initial steps towards living a simpler and fulfilling life. Remember to take it slow and be gentle on yourself as you make the shift. Otherwise, like any crash-diet plan, this shift may not be sustainable either.

FOOD MILES

Food mile is the distance food travels from where it is grown or raised to where it is ultimately purchased by the end user.

Several different methods have been devised over the years to calculate food miles. Here are some of these:

The Weighted Average Source Distance (WASD) formula developed by Annika Carlsson-Kanyama in 1997 is a single distance figure that combines information on distances from producers to consumers and the amount of food product transported.

The Leopold Center for Sustainable Agriculture developed the Weighted Total Source Distance (WTSD) formula, which calculates the weight and distance travelled of each individual ingredient in processed food such as ice cream, pizza, and canned food.

The Weighted Average Emissions Ratio (WAER) formula developed in 2004 by the non-profit Lifecycles takes into account the greenhouse gas emissions associated with each mode of transportation employed for the food to travel from producer to consumer.

It is also important to remember that just ten companies - key among them Nestlé, Mondelez, and Unilever - control almost all large food and beverage companies in the world. Factories are located all over the world, with food and beverages travelling large distances to reach us without us realizing where they are coming from.

Slow Economy

LOOKING AT THE CIRCULAR ECONOMY

'When our wants are moderate, the rations we each claim do not exhaust the common store of nature and the pace of their restoration does not fall hopelessly behind that of our consumption. This moderation leaves us leisure to cultivate happiness, that happiness which is the artist soul of the human world, and which can create beauty of form and rhythm of life....'
RABINDRANATH TAGORE
Introduction to 'Robbery of the Soil' by Leonard Elmhurst

Our relationship with money is broken. It distorts our understanding of life and living, creating a deep dissonance between what **Aristotle** called *chrematistics* – the art of money-making – and *oikonomia*, the art of living, from which the word *economy* originates. This rupture shows up in everyday anxieties: uncertainty about retirement, inadequate healthcare, and worry about securing education for our children in a world shaped by systemic illness rather than well-being.

This dissonance has surfaced in our work. While we were making a documentary on the relationship between the city and its citizens following the privatization of water in **New Delhi** in 2007, we asked people a simple question: *Dilli kiski?* – Whose Delhi is it? In **Khan Market**, a cobbler repairing shoes said, 'I don't know how long I will remain a cobbler; with these disposable PVC shoes that everyone wears now, I too have become disposable.'

His words revealed more than uncertainty. They pointed to an economy built on disposability – one that does not stop with products, but extends to skills, livelihoods, and lives.

A global economy thriving on the illusion of limitless growth develops a limitless appetite for the Earth's resources. This appetite translates into repeated violations of planetary boundaries and the erosion of ecosystems and species. We forget that we are members of an Earth family, and that the real economy is an economy of care – care for the planet and for one another. When care is removed from economic life, insecurity becomes normal, communities fray, and both people and ecosystems are diminished.

The money marathon many of us feel caught in becomes clearer when we examine economic systems shaped by colonialism, industrialization, and globalization. These linear, extractive models rupture nature's cycles of renewal, producing waste, pollution, and social breakdown. Even care – for land, seeds, and communities – becomes commodified, eroding the ethics of reciprocity and interdependence. Over time, this extractive economy encloses what was once held in common. Land, water, seeds, biodiversity, living organisms, and data drawn from our minds and bodies are reduced to private property under the banner of 'intellectual property.' What is lost is not only access, but belonging.

In the chapter on *Slow Economy*, we turn toward another possibility: living economies rooted in *oikonomia* – shaped by circularity, solidarity, mutuality, and cooperation. The art of living lies in recognising and nurturing the regenerative spirit of life – of all beings on Earth, including ourselves.

Slow Food

'Let your place of drinking water be common and your food be together as I yoke you to a common yoke; worship Agni, all of you together, just as the spokes are attached to the navel of the chariot from all around.'
ATHARVA VEDA

Food is the first energy, the currency of the web of life through nutrient cycles. When we abandon the food and energy cycles of the biosphere, when industrial agriculture and globalized, industrial food systems become increasingly dependent on fossil fuels, they contribute to 50 percent of the greenhouse gases. We threaten the regeneration and balance of the living carbon cycle by replacing it with a linear, lifeless carbon pathway that ends quickly and abruptly.

'Agriculture' comes from a combination of the Latin words agrum (form of 'ager', meaning field, farm, land, estate) and cultura (care, growing, cultivation).

Care takes time; carelessness speeds up activities but also destroys. Industrial agriculture that relies heavily on fossil fuels

and chemicals views time and care as obstacles. It has promoted technologies of carelessness by claiming they were 'saving time'. Instead of letting crops ripen with time, Roundup has been promoted as a desiccant and herbicide for instant drying. Machines based on fossil fuels have been introduced to speed up farm operations, destroying biodiversity, livelihoods, and economies of care.

We sped up the growing of food by spraying poisonous chemicals, killing insects and birds. Spreading cancer. We sped up seed breeding by shooting them with gene guns, destroying biodiversity, creating seed monopolies, negating the intelligence of nature evolved over billions of years, and the indigenous knowledge and intelligence of farmers evolved over thousands of years. We sped up food processing to make 'fast food' and 'junk food', creating unemployment and the spread of obesity, diabetes, and other chronic diseases.

Subsequent chapters will detail how one can build food sovereignty and resilience by eating local, artisanal, organic, and healthy as well as how one can be more proactive in growing one's own vegetable and kitchen gardens.

Slow Clothing

'The message of the spinning wheel is much wider than its circumference. Its message is one of simplicity, service of mankind, living so as not to hurt others, creating an indissoluble bond between the rich and the poor, capital and labour, the prince and the peasant.'
MAHATMA GANDHI
Young India

Prior to the Industrial Revolution, fabrics were woven locally with diverse fibres. People would buy durable clothing that could serve them for a long time or make their own from the textiles and resources available to them. Clothing reflected the place, the climate, the biodiversity, and the culture of the people wearing them. We sped up clothing. Today, we have fast fashion, which is pushing farmers to suicide, exploiting workers, and creating huge piles of plastic waste and landfills.

Fast fashion can be defined as mass-manufactured, cheap clothing, which mimics and recreates ideas from fashion weeks, runways and celebrity culture, and turns them into garments in high street stores at breakneck speed. This kind of fashion spells disaster

for the global ecology, while the people producing this fashion are often exploited under the worst working conditions and with poor wages. The consumers of this fashion are cut off from the rich heritage, diversity, and knowledge of human artistry. Thus, we need slow clothing and **fibres of freedom,** which weave non-violence in our lives and show care towards the Earth family from the seed to the farm, and to the spinning and weaving and dyeing of cloth, in order to overcome the environmental and human disaster that fast fashion is actively engaging in.

When we choose 'slow', we choose quality over quantity. We choose humane production systems, which give fair wages to the workers from farm to cupboard. This is what Mahatma Gandhi did when he took up the spinning wheel. This was not only a symbol of the Indian freedom struggle but also a symbol for self-reliance, slow meditation and reflective action, and dignity of labour, which challenged colonialism and has the ability now to challenge neo-colonialism. In the chapter on slow fashion we will look at the true cost of fashion and the role we play in it as well as discuss how to make a shift towards slow, and thus, sustainable fashion.

Slow Health

'When health is absent, wisdom cannot reveal itself, art cannot manifest, strength cannot fight, wealth becomes useless, and intelligence cannot be applied.'
HEROPHILUS
Greek physician, one of the early anatomists

Rapid urbanization has disconnected us from the earth. Our knowledge of elements – which make up the human body, such as air, water, and soil – is also getting eroded as we perennially reach for instant solutions to daily chores, especially in food choices, often due to lack of time. This has led to a hidden pandemic that the medical profession is calling 'lifestyle diseases'. The government and policymakers often say that the public lifestyle problems, such as obesity (along with diabetes that is often associated with it) are their fault: just be a bit more virtuous and you won't get a modern disease. This victim-blaming tends to absolve governments, health bodies, and big businesses from any blame for the modern chronic disease epidemic.

Many medical practitioners have highlighted the 'anthropogenic' factors impacting our health; key amongst them are:

1. NUTRITIONAL VALUE OF WHAT WE EAT

This is a major voluntary factor considered in diseases. Does the food we consume through its hyper-processing have too much fat, sugar, salt, too little fibre, or diversity of minerals and absorbable nutrients? There is obviously a lot to be gained from a good diet, so instead of calorie counting as a norm, we must consider the nutritional value of the food for good health.

2. ACTIVITY (OR MORE CORRECTLY, INACTIVITY)

It is one of the most underestimated disease determinants of modern times. Together, poor nutrition and inactivity are also influenced by and have an influence on stress, leading to health problems such as anxiety and depression.

3. TECHNO-PATHOLOGY OR THE ILL-HEALTH EFFECTS OF TECHNOLOGY

These include injuries from machines and technology such as motor vehicle injuries and warfare. It also includes more obscure problems such as repetitive strain injury due to posture and 'screen dermatology' (skin problems from the electrical output of small screens). Inadequate sleep induced by excess screen time can lead to technology-based injuries such as vehicular and workplace injuries, apart from being an underlying risk for other diseases.

4. THE ENVIRONMENT

Pollution in ambient air, particularly from endocrine-disrupting chemicals, adversely affects human health and accelerates the onset of so-called lifestyle diseases, including cancer and hyperthyroidism.

5. MEANINGLESSNESS, ALIENATION, LOSS OF CULTURE AND IDENTITY

We are just beginning to understand these factors, which relate to the loss of purpose of many in an impersonalized economic system and the displacement of indigenous and other populations. The rising mental health crisis and its relationship to physical health are unfolding at an alarming rate. *Occupation influences health* through biological (shiftwork), physical (exposure), and psychological (stress, peer pressure, conflict) factors.

Drugs, smoking, and alcohol are no-brainers, and one should also include doctor-induced disease from prescribed medications.

Social inequality is also only now being recognized, with many reports showing the significance of relative wealth in a country on health and other social factors.

Slow health acknowledges that these anthropogenic reasons have to be examined and addressed by individuals and communities. In the chapter on slow health, we explore this crisis in more detail and delve into the possible antidotes.

Slow Knowledge

'If you think about it, there are only two people on earth that need to know a lot about remote sensing technology – Monsanto and the CIA.'

JOHN HAMER

Managing Director, Monsanto Growth Ventures

In the first wave of colonialism, colonizer-imposed religion on the natives or 'barbarians' was central to the 'civilizing mission'. Today's neo-colonialism, imposing the dogma of fast, centralized, and all-pervading technology on our diverse food systems and diverse knowledge and technologies systems on which they are based, is central to the new civilizing mission.

When technology is no longer seen as a tool to be assessed, chosen, adopted or rejected, but as a religion, as a civilizing mission that is to be forced undemocratically on people, and when means for money-making are elevated beyond ethical, social, ecological and democratic assessment, we are subjected to colonization, only this time in a modern garb.

If we shape new technologies and use them as instruments and

tools, we can take their help to facilitate our understanding; for example, indigenous farmers know the soil is living, but it takes a microscope to see the microbes in the soil food web. However, if the tools of the powerful are used as instruments that deepen our separation and reduce most people's power, and put it in the hands of a few, then this crisis is aggravated.

In the case of agriculture, such technologies will take us further down the road of separation and eco-apartheid, further denying the intelligence of farmers and of plants, besides perpetuating a culture where the farmers' traditional knowledge systems of plants, nourishing the soil, and water retention are made obsolete.

Farmers have precious experiential knowledge that is being harvested through digitalization. But data is not knowledge. It is just another commodity to make the farmer more dependent. The farmer is being told that they must outsource their mind to big agrotech industry. This is the next step in a dead-end future that ignores the intelligence of seeds, plants, soil organisms, our gut bacteria, our farmers, and our grandmothers. We have already experienced its great danger during the much-celebrated Green Revolution. These companies tried to steal the knowledge of India's ancestors through biopiracy and patent laws, and today, they have gone a step further – biopiracy is carried out through the convergence of information technology and biotechnology.

Gyan Swaraj, or 'knowledge sovereignty', is the cornerstone of slow knowledge. In the chapter on slow knowledge, we explore how we can maintain knowledge sovereignty and cultivate slow knowledge so that a few corporations don't end up deciding every aspect of our lives. Knowledge diversity is the celebration of our rich heritage, wisdom, literature, art, poetry, and *who we are* as a human collective.

Slow Democracy

'For when people do not keep watch over the commons, it is destroyed. It results, then, that they fall into civil faction, compelling one another by force and not wishing to do what is just themselves.'
ARISTOTLE
The Nicomachean Ethics

Slow democracy is living democracy. It puts life and freedom of all life at the centre of democracy. When we think of democracy today, it is as a political system of occasionally participating in voting for a representative and reneging on our responsibilities to them until the next voting cycle. Transparency, accountability, and the practices of democracy, however, are much deeper than that. It's an act of discerning action, dialogue, and understanding of power that has to take place every day, slowly and constantly. It is the recognition that from the rights of all beings flow human responsibilities to care for the Earth and all its beings, including fellow human beings.

Slow democracy is, therefore, the practice of compassion in our daily lives. Slow democracy is the practice of sharing the Earth's gifts, her ecological space, with all beings and using the Earth's resources within ecological limits and planetary boundaries. It is 'commoning' – the reclaiming and creation of the Commons. Commons and community go hand in hand. Slow democracy is based on participation in decisions of everyday life. Participation is not reducible to representation.

The first principle of the Earth Democracy movement is waking up to our being part of nature and one with the Earth. We call the illusion of separation from the Earth 'eco-apartheid'. This gives rise to anthropocentrism, the illusion of humans being superior to other species, whom we can manipulate as objects and push to extinction. These illusions have created economies that kill both nature and people. Movements are increasingly referring to ecocide and genocide in the economic context. Such destructive economies also destroy democracy and take away our freedoms, in their wake creating cultures of hate and fear.

The shift to Earth democracy as a worldview allows us to imagine and grow living economies, living democracies, and living cultures. The chapter on slow democracy looks at some oases of slow democracy and inspires us to create our own spaces for practising and living in slow democracies.

Slow Culture

**'Life in the fast lane surely makes you lose your mind.
Eager for action and hot for the game.
The coming attraction, the drop of a name.
They knew all the right people.
They took all the right pills.
They threw outrageous parties.
They paid heavenly bills.
There were lines on the mirror, lines on her face.
She pretended not to notice she was caught up in the race.'**
'LIFE IN THE FAST LANE', THE EAGLES
Hotel California (1976)

Slow culture is an invitation to look within and examine our choices and the kind of future we want to build. It's an evolving discourse stemming from the understanding that the fast-paced life is creating paradoxes and conflicts that can be resolved only if we slow down and listen. Listen to ourselves, to the connections between animate and inanimate objects, and take the time to build societies that are inclusive, diverse, egalitarian, non-violent, and democratic. It is a call to learn from Mother Earth and imbibe a sense of universal responsibility and care. As Mahatma Gandhi said, action without reflection is just reaction; thus, slow culture asks us to become more mindful and reflective in our journey in life and to think of the consequences of the way we consume, the way we relax, engage in leisure, and celebrate life. After all, how we travel is as important as the destination itself. It is critical to engage in understanding and practising equality, interbeing, and non-violence.

READING LIST

VANDANA SHIVA, *Earth Democracy: Justice, Sustainability, and Peace.* South End Press, 2005.

M.K. GANDHI, *Hind Swaraj*. Delhi Open Books, 2021.

E.F. SCHUMACHER, *Small is Beautiful: A Study of Economics as if People Mattered.* Vintage Books, 1993.

ERIC FROMM, *The Art of Loving*. Harper, 1956.

DUANE ELGIN, *Voluntary Simplicity: Toward a Way of Life That Is Outwardly Simple, Inwardly Rich.* Harper Collins, 1993.

WOODY TASCH, *Inquiries into the Nature of Slow Money: Investing as if Food, Farm and Fertility Mattered.* Chelsea Green Publishing, 2010.

Food to Slow Down and Reset

STAY IN BALANCE *with* KHICHDI

INGREDIENTS

1 cup rice
1½ cup green gram *(sabut moong)* or red gram *(masoor dal)*
3 cups water
Salt, to taste

For seasoning
2 tbsp ghee or any other cooking fat
2 cloves
1 cm piece cinnamon
1 pinch turmeric powder

A one-stop bowl complete meal, khichdi aids in digestion and is the best anti-inflammatory food.

METHOD

- Wash the rice and dal together, and drain.
- Heat the ghee or oil in a pot over medium heat and prepare a seasoning adding the cloves, cinnamon, and turmeric.
- Add the rice and dal to the pot and stir for a few minutes.
- Add water and season to taste with salt.
- Cook over low heat until steamy, or in a pressure cooker for up to four whistles.
- For variations, the seasoning can be made with mustard seeds (2 teaspoon), green chillies (two, slit length-wise in the middle), a pinch of asafoetida, and turmeric (1 teaspoon), or with mustard seeds (2 teaspoon) and green chillies (2, slit in the middle) only.

Tip

- For more taste variation, you can add vegetables like French beans, green peas, potatoes, cauliflower, and carrots.

Source: Akshat, *Navdanya*

REPAIR AND REJUVENATE *with* RASAM

INGREDIENTS

4 tbsp pigeon peas
1 tsp tamarind paste
3 cups water
Salt, to taste
1 tsp grated jaggery or cane sugar
2 tbsp sunflower oil or any cold-pressed oil
1 tsp black mustard seeds
1 tsp cumin seeds
Large pinch of asafoetida
10 curry leaves
4 cloves garlic, peeled
1 tsp sambar powder
1 tsp turmeric powder
1 tsp red chilli powder
Fresh coriander for garnish

Rich in minerals, rasam helps rejuvenate skin, and is also effective in weight loss.

METHOD

- Wash the pigeon peas (tuar dal) and soak in water for 5 minutes and drain.
- Dilute the tamarind pulp in 2 tablespoons of water and keep aside.
- In a pan over medium heat, add the soaked lentils and 3 cups of water.
- Simmer until the dal is soft. Remove from the heat and keep aside.
- Season to taste with salt and add the jaggery, stirring until it dissolves completely.
- Heat the oil in a deep pan over medium heat.
- Add the mustard seeds and let them sputter. Add the cumin seeds, asafoetida, curry leaves, and garlic cloves. Sauté for a minute.
- Now add the diluted tamarind and simmer until it becomes slightly thick and bubbly. Add the sambar powder, turmeric, and red chilli powder.
- Pour the prepared lentils into the tamarind mixture and bring to a boil. Remove from the heat immediately and serve hot, garnished with fresh coriander leaves.

Source: Masale – Spices, *Navdanya*

RELAX AND RECUPERATE *with* SAFFRON MILK

INGREDIENTS

1 cup full-fat, preferably organic milk

5–7 almonds, soaked & peeled

1 or 2 dates

1 tsp ghee or coconut oil

A few threads of saffron pre-soaked in 1 tsp of milk for 30 minutes

A pinch each of cardamom, cinnamon, and nutmeg powder

METHOD

- Blitz all the ingredients together in a blender until smooth and frothing.
- Transfer into a saucepan over medium heat and bring to a boil.
- Serve warm an hour before bedtime.

An excellent detoxifier, it relieves coughs and colds and combats insomnia and depression. It also helps ease severe menstrual cramps.

Source: Biodiversity in a Glass, *Navdanya*

2

COUNTERING *the* CLIMATE CRISIS

Conserving *and* Restoring *our* Relationship with Earth

The Earth is living and life-giving. For over four billion years, the Earth has evolved a rich biodiversity – an abundance of different living organisms and ecosystems – that can meet all our needs and sustain life. All organisms and their inorganic surroundings on Earth are closely integrated to form a single and self-regulating complex system, maintaining the conditions for life on the planet. Introduced in the early 1970s, the idea was conceived by chemist and inventor James E. Lovelock and biologist Lynn Margulis. They called it the Gaia hypothesis, named after the ancient Greek goddess of the Earth. It postulates that the Earth and its biological systems behave as a huge single entity. As the two studied the processes by which living organisms produce and remove gases from the atmosphere, they realized that the Earth is a self-regulating living organism, and life on Earth creates conditions for life to be maintained and evolve. This hypothesis was a scientific reawakening to the living Earth.

This crisis of planetary imbalance that we are facing today is a consequence of nearly two centuries of dependence by industrialized countries on fossil fuels. Industrialization, in fact, is equated with a transition to fossil-fuel economy. Unfortunately, development too has been defined as industrialization. The earth fossilized living carbon and transformed it into dead carbon, storing it underground; that is where we should have left it. Instead of building civilizations around dead carbon, we must become in sync with the earth and turn to living carbon, such as organic manure, to build our world. We did this in the past and the balance was maintained. As we broke our contract with the earth and started to view ourselves as separate, we face an auto correction through living earth – climate crisis.

In biology, the term development refers to self-directed, self-regulated, and self-organized evolution from within. In the terms of Chilean biologists Humberto Maturana and Francisco Varela, systems that self-organize and self-renew are autopoietic systems. If the economic domain were to think of development in the same way, it would lead to the flourishing of biodiversity and cultural diversity. Development would conserve resources and energy while improving well-being and human welfare.

A few centuries of fossil fuel-based civilization have brought our very survival under threat by rupturing the earth's carbon cycle, disrupting key climate systems and its self-regulatory capacity, and pushing diverse species to extinction at 1,000 times the normal rate. The connection between biodiversity and climate change is intimate. Take the example of food and agriculture systems. The earth has roughly 300,000 edible plant species, but the contemporary global human community eats only 200 of them. And, according to the *New Scientist*, 'half our plant-sourced protein and calories come from just three: maize, rice and wheat.' Meanwhile, only 10 percent of the soy that is grown is used as food for humans. The rest goes to produce biofuels and animal feed.

We have been repeatedly told that monocultures of crops with intensive chemical inputs of synthetic fertilizers, pesticides, and herbicides are necessary for feeding the world. While using 75 percent of the total land that is being used for agriculture, industrial agriculture – based on fossil fuel-intensive, chemical-intensive monocultures – produces only 30 percent of the food we eat; small biodiverse farms using 25 percent of the land provide 70 percent of the food. Industrial agriculture is responsible for 75 percent of the destruction of soil, water, and biodiversity of the planet. At this rate, if the share of fossil fuel-based industrial agriculture and industrial food in our diet is increased to 40 percent, very soon we will have a dead planet with no life and no food.

Besides the carbon dioxide directly emitted from fossil fuel-based agriculture, nitrous oxide is emitted from nitrogen fertilizers, and methane is emitted from factory farms and food waste. The manufacture of synthetic fertilizer is highly energy-intensive. One kilogram of nitrogen fertilizer requires the energy equivalent of two litres of diesel. The total energy used during fertilizer manufacture was equivalent to 191 billion litres of diesel in 2000 and is projected to rise to 277 billion litres by 2030. This is a major contributor to climate change, yet largely ignored. One kilogram of phosphate fertilizer requires half a litre of diesel. Nitrous oxide is 300 times more disruptive to the climate than carbon dioxide. Nitrogen fertilizers are destabilizing the climate, creating dead zones in the oceans and desertifying soils. In the planetary context, the erosion of biodiversity and the transgression of the nitrogen boundary are serious, though often-overlooked, crises.

Organic farming takes excess carbon dioxide from the atmosphere and puts it back into the soil where it belongs, through photosynthesis. It also increases the water-holding capacity of soil, contributing to resilience in times of more frequent droughts, floods, and other climate extremes. Organic farming has the potential to sequester 52 gigatons of carbon dioxide, equivalent to the amount needed to be removed from the atmosphere to keep atmospheric carbon below 350 parts per million and the average temperature increase of two degrees centigrade. We can bridge the emissions gap through supporting and undertaking ecological biodiversity-intensive agriculture, working with nature. By doing this, we will start to repair the broken carbon cycle.

Dead Carbon Lifestyle vs. Living Carbon Lifestyle

Maati kahe kumhaar se
Tu kya rundhe mohe
Ek din aisa aayega
Main rundhungi tohe

The earth says to the potter
You think you are shaping me?
There will come a day
When I will shape you
KABIR DAS (1440–1518)

When we disrupt the Earth's self-regulating climate systems, we get climate chaos and climate catastrophe, which we can 'geo-engineer' or 'genetically engineer' our way out of. Living systems evolve, adapt, and regenerate; they are not engineered.

DEAD CARBON	LIVING CARBON
Extractive economies – mining earth's resources using heavy machinery and fossil fuel-based economies.	**Rejuvenating economies** tilling the earth, composting, and rejuvenating oceans, rewilding forests and skies.
Turning labour into a commodity through mass production, reducing our creativity.	Artisanal production giving wings to creativity and dignity of labour.
Defining development as an aspiration for urban living, leading to urban migration and unsustainable megacities.	Sustainable living in villages, cities, and towns based on a deep connection to the earth through community gardening, farmers' markets, and renewable energy.
Gambling our future away in carbon casinos by trading mythical solutions through carbon trading.	Caring for seven generations by implementing 'real' solutions to climate change, such as water harvesting, urban gardening, introducing a sustainable public transport system, and taking action against mega polluters.

Carbon Trading

CARBON CASINO: GAMBLING THE EARTH'S FUTURE

'When the last tree is cut down, the last fish eaten and the last stream poisoned, you will realize that you cannot eat money.'
NATIVE AMERICAN SAYING

According to Fern, an environmental and social justice organization, 'carbon trading is the process of buying and selling permits and credits to emit carbon dioxide.' This approach suggests that the Earth's capacity to regulate its climate can be treated as a measurable commodity. Today, carbon trading lies at the centre of global climate policy and is projected to become one of the world's largest commodities markets. However, it has a disastrous track record since its adoption as part of the Kyoto Protocol.

In their booklet *Carbon Trading: How It Works and Why It Fails*, the Transnational Institute (TNI), an international non-profit research and advocacy organization, highlights that under a scheme called 'cap and trade', governments or intergovernmental bodies like

the European Commission hand out licenses to pollute (or 'carbon permits') to major industries. Instead of cleaning up their act, one polluter can then trade these permits with another who might make changes more cheaply for the equivalent amount. This is the approach underlying the European Union's Emissions Trading Scheme (EU ETS), the world's largest carbon market, which is currently worth €683 billion (approximately $769 billion). Thus, carbon trading transforms the ethical and ecological problem of reducing our carbon footprint into a financial opportunity to make billions.

The second type of carbon trading is offsetting. Instead of cutting emissions at source, companies, and sometimes international financial institutions, governments, and individuals, finance 'emissions-saving projects' outside the capped area. Although offsets are often presented as emissions reductions, they do not actually reduce emissions. Even in theory, at most they merely move 'reductions' to where it is cheapest to make them, which normally means a shift from Northern to Southern countries. Pollution continues at one location on the assumption that an equivalent emissions saving will happen elsewhere. The projects that count as 'emissions savings' range from building hydroelectric dams to capturing methane from industrial livestock facilities. According to Dan Welch, a clean mobility expert and researcher at the WRI Ross Center for Sustainable Cities, 'Offsets are an imaginary commodity created by deducting what you hope happens from what you guess would have happened.'

Carbon trading allows polluters to keep polluting by buying credits. The credits do not value the least polluting activity or pathway available, but instead, reward a tiny reduction in heavily polluting activity. These credits can, in turn, be sold to buyers who can then use them to pollute over their allocation. Factories and plants that exceed their allocation are required to buy credits but are not required to cut their pollution. According to the Transnational Institute, 'the theory is that the availability of carbon permits will gradually be reduced, ensuring scarcity, so that the market retains its value while at the same time forcing a reduction in the overall level of pollution.'

Carbon trading has created a system whereby different greenhouse gases are treated as equivalent and quantifiable 'things', opening them up to the possibility of exchange. While carbon trading encourages ingenuity in inventing measurable 'equivalences' between emissions of different types in different places, it does not select for innovations that can initiate or sustain a real shift away from fossil fuels. For both governments and many large corporations, the appeal of carbon trading schemes is that they give the appearance of addressing climate change but do not mandate an immediate start to structural change in existing energy practice. To build a truly sustainable world we have to move from oil to soil. This carbon casino of imaginary exchange is a mirage and not a solution. It shows a lack of courage towards systemic change as carbon-addicted civilizations willingly hurtle towards the destruction of the planet.

What is needed is a carbon tax on corporations – both for their production system, no matter where their facilities are located, and for transport. A system which subsidises long-distance transport while local carbon emissions production and distribution are penalized and high emission of long-distance transport is excluded, is counter-intuitive. It assures that dirty industries continue to prosper merely by shifting them from one place to another. But we forget the atmosphere, oceans, and air belong to one Earth that has no national boundaries, and thus the impact on the collective is the same. Relocating the polluters is not a solution. Shifting from dead carbon to living carbon is our only hope.

Conscious Consumer

5 SIMPLE STEPS TO REJECT A DEAD-CARBON LIFESTYLE

'Every time you spend money, you're casting a vote for the kind of world you want.'
ANNA LAPPÉ
Diet for a Hot Planet: The Climate Crisis at the End of Your Fork and What You Can Do About It

1. EAT AND WEAR BIODIVERSE AND CLIMATE RESILIENCE

In the 1970s, the World Bank gave massive loans to India to promote groundwater mining. It forced states like Maharashtra to stop growing water-prudent millets like jowar, which needed only 300 mm of water, and shift to water-guzzling crops like sugarcane, which required 2,500 mm of water. In a region with 600 mm rainfall and 10 percent groundwater recharge, this was a recipe for water famine (see Navdanya's *Financing the Water Crisis*). Not only has water-wasteful chemical agriculture mined groundwater, it has also mined soil fertility and contributed to climate change. Chemical fertilizers destroy the living processes of the soil and make it more vulnerable to drought. Chemical fertilizers also produce nitrous oxide, a greenhouse gas 300 times more potent than carbon dioxide. The solution for the climate crisis, the food crisis, or the water crisis, under which India is reeling, **is to return to** biodiversity-based organic farming systems. Biodiverse ecological farms address the climate crisis by reducing emissions of greenhouse gases such as nitrous oxide and absorbing carbon dioxide in plants and soil. Biodiversity and soils are the most effective carbon sinks. They also help adapt to climate change and drought by increasing soil organic matter, which increases the moisture-holding capacity of soil, and hence provides drought-proofing of agriculture. Biodiverse organic farms increase food security by increasing the resilience and reducing the climate vulnerability of farming systems. They also enhance food security because they have a higher production of food and nutrition per acre than Green Revolution monocultures, which measure the yield of our commodity, not the total food output or the nutritional quality of food.

Biodiverse organic systems also address the water crisis. Firstly, production based on water-prudent crops like millets reduces the demand for water. Secondly, organic systems use ten times less water than chemical systems. Thirdly, by transforming the soil into a water reservoir through increasing its organic matter content, biodiverse organic systems reduce irrigation demand and help conserve water. Maximizing biodiversity and organic matter production thus simultaneously increases climate resilience, food security, and water security. If we eat, wear, and use products made out of hardy crops like jute, hemp, and millets, we can increase the demand for this biodiversity in the market and thus help shift the cultivation pattern from monoculture to biodiverse agriculture.

2. EAT LOCAL AND WEAR LOCAL

Local food produce is fresh, healthy, and tastes better because it spends less time in transit from farm to plate, and therefore loses fewer nutrients and incurs less spoilage. Local food encourages diversification of local agriculture, which reduces the reliance on monoculture – single crops grown over a wide area to the detriment of soil health. Thus, it conserves local biodiversity and encourages you to eat seasonal and diverse food. It reduces the number of preservatives and packaging required to increase its shelf life as well as helps stimulate the local economy.

HEALTH PER ACRE	CALORIES PER ACRE
Diversity Organic biodiversity-based mixed cropping is the foundation of the concept of health per acre.	**Monoculture** A few food items produced abundantly cannot ensure an ideal blend of nutrients supplied to every person, as any single food item is not the adequate source of all nutrients needed by the human body. The question we need to ask is if abundant production of rice, wheat, corn, or soybean would solve the crisis of undernutrition or if abundant production of all the different nutrients would.
Organic farming People and environment-centric agriculture which does not poison the seed, soil or water and uses carbon sequestering practices for fertilization ensures better health both for the producer and the consumer, and for the environment.	**Poison cartel** Agro-industry-based profit-driven food which disempowers small-household farmers, making them market dependent on terminator seeds (seeds that are not progenitor). High use of chemical fertilizers and pesticides ensure that the food that we eat will make us sick in the long run as well as contaminate the environment, hurting biodiversity in and around the farm.

While looking at consuming local from a food miles and clothes mile perspective, keep in mind that the mode of transport to your destination makes a difference to the equation. For example, flying one tonne of food is nearly 70 times more carbon-intensive than transporting that same weight via a large cargo ship or transporting it by train to your destination. Looking exclusively at carbon footprints through the lens of transportation also neglects other important issues like water usage, the type of farming practices used, and farmer rights. It is therefore important for consumers to define the values you hope to support through your purchasing decisions and think critically about when and where local foods and clothes support those values. Thus, ticking just one box may be counter-intuitive to what you are trying to achieve. Wearing local fibres does the same for farmers and you. It reduces the energy consumed in transportation, as the journey of access is shorter and thus less energy-consuming. We must think like the living Earth and in terms of the web of life rather than fragmented one-off solutions. It is only then that our actions will stand up to scrutiny.

3. EAT AND WEAR ORGANIC AND POISON-FREE

The International Assessment of Agriculture Knowledge, Science and Technology for Development (IAASTD) Synthesis Report was the largest review of our current global agricultural system ever undertaken. This multi-stakeholder process involved over 400 scientific authors and 61 countries. According to the report, industrial agriculture contributes about 60 percent of anthropogenic emissions of methane and about 50 percent of nitrous oxide emissions. Improper fertilization has caused eutrophication and large dead zones in several coastal regions. The use of pesticides that are unsuitable has led to groundwater pollution, health problems, and loss of biodiversity. The IAASTD report did not endorse the GMO and industrial agriculture paradigms as the solution to improve sustainability and instead proposed working at a more local level with lower inputs, family farms, and endorsed ecological farming methods, including organic farming.

Industrial monoculture agriculture has pushed more than 75 percent agro-biodiversity to extinction. Nearly 75 percent bees have been killed by toxic pesticides. Einstein famously warned us that **when the last bee disappears, humans will disappear too.** Thus, if we calculate the true cost of non-organic food on health and the planet, we will realize it will cost us and the planet much more in the long run than the perceived high cost that we feel while buying organic. Furthermore, the more organic we eat, the more likely the cost of organic produce will go down, as demand and supply will be in sync with each other. We will explore in detail the health costs and true costs of poison in our food in the chapter on slow health. A similar principle applies to the clothes we wear, for the use of plastic in our clothes is astounding. As we know, plastic is manufactured using petrochemicals and has been filling up our landfills for aeons. The chapter on slow fashion will help you understand how to spot plastic in your clothes so you can avoid consuming it.

4. BUY ARTISANAL AND HANDMADE

The process of production is equally important for understanding the impact of the product on the carbon cycle. The more mechanical and factory-made the product, the more the use of fossil fuel in the production of the goods you are consuming. Traditional handicrafts are also a storehouse of conventional wisdom through which we can learn and promote a chemical-free lifestyle, such as natural dyes for clothes; the use of jute, bamboo, and other eco-friendly materials to make office stationery; and the use of stone, clay, and hand-beaten metal for cutlery and cooking. All this helps not just preserve local knowledge but also enables local craftspeople and artisans and, above all, contributes towards a less energy-intensive and decentralized production system that can help us combat the climate crisis and uphold diversity.

5. LIVE PLASTIC-FREE

While plastic pollution in the oceans has become a burning concern, the actual effects on climate change of the ubiquitous use of plastic has not been focussed on. Almost all plastic is derived from materials (like ethylene and propylene) made from fossil fuels (mostly oil and gas). Emissions from plastic emerge not only from the production and manufacture of plastic itself but from every stage of its lifecycle – from the extraction and transport of the fossil fuels that are the primary feedstock for plastic to refining and manufacturing to waste management. For example, 4 percent of the world's annual petroleum production is diverted to making plastic, and another 4 percent gets burnt in the refining process.

The contribution of plastic production and disposal to climate change has been largely hidden, says the report by the Centre for International Environmental Law, which estimates the greenhouse gas footprint of plastic from the cradle to the grave for the first time. According to this report published in 2019, the lifecycle of global plastic production – from extraction to disposal – was equivalent to the impact on the climate of 189 500 MW coal-fired power stations. The report predicts that by 2050 the global plastic footprint will be equivalent to 615 coal plants running at full capacity. Apart from that, the disposable plastic found in packaging and fast-moving consumer goods forms the largest and fastest-growing segment of the plastic economy.

The report further states throwaway plastic packaging makes up 40 percent of its demand, fuelling a boom in production from 2 million tonnes in the 1950s to 380 million tonnes in 2015. By the end of 2015, 8.3 billion metric tonnes of plastic had been produced – two-thirds of which had been released into the environment and remains there. Thus, getting rid of plastic from our daily life is survival-imperative for the climate. This doesn't mean we throw out the existing plastic in our life, for that will only become trash to pollute the ocean. Instead, we must find ways to reuse and recycle existing plastic in our homes and not buy anything new. This includes shopping with an awareness and consciousness that rejects plastic packaging, trying to reduce takeaway food in plastic containers, taking instead our own tiffin boxes to restaurants for takeaway, and carrying our own cutlery and straws, mugs for coffee, and bottles for water.

Repairing the Broken Carbon Cycle

'The soil is the great connector of lives, the source and destination of all. It is the healer and restorer and resurrector...'
WENDELL BERRY
The Unsettling of America: Culture & Agriculture

Living carbon sustains life on earth. However, we have broken this cycle by becoming a fossil fuel-dependent society. To repair the broken carbon cycle, we need to turn to seeds, to the soil, and to the sun to increase the living carbon in our plants and soil. Dead carbon must be left underground – an ethical obligation and an ecological imperative. The term 'decarbonization', which fails to make a distinction between living and dead carbon, is scientifically and ecologically inappropriate. If we decarbonized the economy, we would have no plants or living carbon, and thus no life on earth.

The more we grow biodiversity and biomass in forests and farms, the more organic matter is available to return to the soil, thus reversing the trends towards desertification. Unfortunately, desertification is already a major reason for the displacement and

uprooting of people and the creation of refugees in sub-Saharan Africa and the Middle East.

Biodiversity-based agriculture is not just a climate solution. It is also a solution to hunger. Approximately one billion people are permanently hungry in the world. Systems that are biodiversity-intensive, fossil fuel-free, and chemical-free produce more nutrition per acre and can feed more people using less land. Let us invite life back into our lives and move away from processed food to nutritious food in our daily meals. Support and plant a harvest of hope and become a society that embraces living carbon.

REPAIRING THE CARBON CYCLE AND RESPONDING TO THE CLIMATE CRISIS: THE NAVDANYA WAY

The Navdanya Movement was established by Dr Vandana Shiva in 1987. This journey of practice and churning through the various Navdanya initiatives has validated our hypothesis of living carbon. It has taught us that we don't grow food, we grow biodiversity, pollinators, microorganisms, and soil fertility. They in return give us a third of our food; microorganisms increase the soil productivity and bless us with the gifts of abundance. Our agro-ecological practices and organic food systems have regenerated the groundwater table from 120 feet to 60 feet in the last two decades. Navdanya has established over 124 seed banks in 22 states, trained over one million farmers and helped transition two million acres to organic farms in India alone. Since its conception, Navdanya has been cultivating dynamic earth communities and are co-evolving the web of life. The community seed banks have helped farmers move from vulnerability to resilience.

Averting the cyclone and drought crisis in Odisha, Navdanya has co-evolved 813 paddy varieties; which include 58 drought-resistant, 50 flood-resistant, 33 saline-resistant, 2 therapeutic, and 55 aromatic rice varieties. The community saved varieties like 'Bhundi Kambanh', 'Lunabekada', and 'Sankarchin', which have helped rehabilitate agriculture in coastal villages of Tamil Nadu after the 2004 tsunami. Keeping climate resilient seeds in farmers' hands is vital for food security in times of climate chaos. Our Seeds of Freedom are currently being harvested in remote earthquake-affected areas of Nepal. These indigenous seeds have revived agriculture in the region and helped farmers and consumers move towards food sovereignty. Seeds of Freedom and Gardens of Hope (Saagvaris or edible schoolyards) teach our children the sacred traditions of seed saving and organic gardening, with 300 schools all over India. The kitchen garden ensures food and nutrition and is an integral part of repairing the carbon cycle and addressing the climate crisis.

Planning your Garden of Hope

'City wisdom became almost entirely centred on the problems of human relationships, in contrast to the wisdom of any natural tribal group, where relationships with the rest of the animate and inanimate world are still given due place.'
JAMES LOVELOCK
Caia: A New Look at Life on Earth

Restoring our broken relationship with the Earth and all its animate and inanimate beings is crucial to overcome the climate crisis. I have seen from the Navdanya experience how modern agriculture has tried to break our relationship with earth and how our indigenous knowledge has been successfully used to overturn this tide. The wisdom of Navdanya's practices, which reflect the accumulated agricultural knowledge of many generations of farmers, adapted to the modern context of urban agriculture, and the many interns who come have helped us set up Gardens of Hope in different schools and communities. Let us start this journey by imagining our Garden of Hope together to repair the carbon cycle.

This journey is about returning to our roots, to the most important elements in life, such as seeds and soil. It is about taking care of our fellow human beings, and most importantly, **Mother Earth**! If you live in a city, you may feel that there are no opportunities to garden because buildings and roads have replaced our fields and made the production of food an activity that occurs mainly in rural India, hundreds of kilometres away from us.

Contrary to popular belief, cities are full of opportunities to garden, whether it is on rooftops, windowsills, balconies, or, for the luckiest, in a small garden. The practice of urban agriculture has been growing as the many benefits of fresh, locally produced food gain greater recognition.

However, we do not guarantee a successful harvest. Gardening is a learning process, and it is with patience and critical reflection that you will find the techniques that work for your garden. However, we hope that this can act like basic tools to start a Garden of Hope and to deal with common pest control, harvesting, and seed-saving issues.

PLANNING YOUR GARDEN

When starting a garden, it is important to pay attention to these details: the layout, number/type of seeds to be sown, watering amounts, pest and insect remedies, when to harvest and how to save the seeds for future use. All this requires careful planning before starting our journey of actually sowing and planting a garden.

We can begin planning our Garden of Hope in two broad ways:

OPTION ONE

Choosing the spot

If you have limited options to grow crops, explore the opportunities offered to you. For example, if you do not have a garden, growing crops in pots is a great alternative, as you can select crops that are adapted to pot gardens. When choosing a space for your garden, keep in mind how you will water it at the selected location and how much sunlight the location receives. Remember, most plants like the morning sun the best!

Below are some ideas for spaces you can create for your garden. A modest garden that is well taken care of will give a better harvest than a large garden that is difficult to maintain, so in the initial stages you may want to start small and expand as your knowledge, skills, and love for gardening grow.

No book can teach you where to start your urban garden. It is over time, through maybe numerous experiments in different places that you will figure out the best place to grow your plants.

OPTION TWO

Choosing the crop

You can select your crops first. Refer to their needs, and it will help you choose if you need a humid or dry, sunny or shady place to grow your garden.

Figuring out what to plant in your garden is a matter of personal taste. What would you want to eat? Taking into account the growing season, climatic region, and unique environmental conditions of your particular site. In India, there are two main seasons largely taken into account when farming: Rabi, known as the winter season, and Kharif, the monsoon season. The Jaid season, from April to June, links both seasons. To start deciding on the crops you want to grow, refer to the season you want to start sowing.

WHERE CAN YOU DO THIS?	WHAT KIND OF GARDEN CAN YOU DO?	HOW BIG CAN YOUR GARDEN BE?
A playground, a backyard with soil	A garden in a plot of land	$1\ m^2$ is a good size to begin with
Concrete playground, yard, balcony or even roof	Beds or pots	40 cm high!
Not so much space and a cemented floor but containers!	Vertical garden	It will depend on the size of the container you find. Start with a small one!

HOW CAN YOU DO THIS?

Clear the stones, remove any garbage, but leave organic matter (leaves) to build the nutrients in the soil. Check for top soil. If there is not sufficient top soil (10 cm), then add top soil. Churn the soil and ensure the soil is moist to promote seed germination. Take care that this space is safe and that you can protect it from damage by trampling.

Use wood or other natural or reclaimed material to create garden beds. The beds should allow for 15–20 cm of top soil, 10 cm of secondary soil (wood chips, branches, and other organic matter) and 10 cm of stones and sand for water percolation. Check for the structural capacity of the rooftop or the balcony to make sure it can support the added weight of a garden. In many cases, you can build a barrier between the roof and the growing medium to prevent root growth.

Carefully cut holes in the side of the container. Fill the bottom ¼th of the container with stones and sand. Fill the rest of the container with nutrient-rich soil. If your container is larger than 20 litres, it will be helpful to insert vertical bamboo reeds just above the level of the various seed holes until above the top of the container. Water can be poured down the holes to help with proper water distribution. Plant seeds in the holes cut into the container and at the top of the container.

MONTH	COOL & DRY AREAS	WARM & TROPICAL AREAS	HIGH RAINFALL ZONES
January	Brinjal	Lettuce, spinach, gourds, melons, radish, carrot, onion, tomato, okra, brinjal, beans	Spinach, mustard greens, pumpkin, brinjal, ash gourd, flat beans, local leafy greens
February	Apple gourd, bitter gourd, bottle gourd, cucumber, French beans, okra, sponge gourd, watermelon, spinach	Same as January	Bottle gourd, sponge gourd, ridge gourd, okra
March	Same as February	Amaranthus, coriander, gourds, beans, melons, spinach, okra	Ash gourd, ridge gourd, bottle gourd, beans, cucumber, leafy greens, watermelon
April	Capsicum	Onion, amaranth, coriander, gourds, okra, tomato, chilli	Tomato, okra, taro leaves (arbi patta)
May	Onion, pepper, brinjal	Okra, onion, chilli	Chillies, snake gourd, pumpkin
June	All gourds, brinjal, cucumber, cauliflower, okra, onion, sem (runner bean), tomato, pepper	Gourds, solanaceae, almost all vegetables	Water spinach (kalmi saag), pumpkin, sweet potato

MONTH	COOL & DRY AREAS	WARM & TROPICAL AREAS	HIGH RAINFALL ZONES
July	All gourds, cucumber, okra, sem, tomato	Same as June	Same as June
August	Carrot, cauliflower, radish, tomato	Carrot, cauliflower, beans, beet	Same as June
September	Cabbage, carrot, cauliflower, peas, radish, tomato, lettuce	Cauliflower, cucumber, onion, peas, spinach	Same as June
October	Beet, brinjal, cabbage, cauliflower, lettuce, peas, radish, spinach, turnip	Brinjal, cabbage, capsicum, cucumber, beans, peas, spinach, turnip, watermelon	Spinach, fenugreek (methi), mustard greens, radish, carrot, brinjal, tomato, coriander
November	Turnip, tomato, radish, pepper, peas, beet	Beet, eggplant, cabbage, carrot, beans, lettuce, melon, okra, turnip	Same as October
December	Tomato	Lettuce, pumpkin, watermelon, muskmelon, ash gourd, ridge gourd, bitter gourd, bottle gourd, cucumber, chilli, cabbage	Same as October

It is important to understand the particular needs of different plants when planning your garden. Knowing the botanical relationships between different vegetables, herbs, and flowers can help you determine the layout of your garden. In fact, you can practise **intercropping**, where complementary crops can be combined to improve soil fertility and support good yields.

IT IS EASIEST IF PLANTS ARE GROUPED ACCORDING TO their harvesting period. For example, everything that will be harvested early should be in the same area so that when they are all over, you can clear the area and plant something else. It is always a good practice to rotate your vegetable garden planting. This helps to prevent pests from becoming established and returns certain nutrients to the soil.

COMPANION PLANTING CHART FOR VEGETABLES

VEGETABLE	GOOD COMPANIONS	BAD COMPANIONS
Beans	Brinjal, carrot, cabbage, cauliflower, cucumber, celery, corn, marigold. *(Corn protects against wind and sun and provides climbing support. Squash has deep roots, beans are shallow, and squash smothers weeds and provides a living mulch)*; cucumber, strawberries *(particularly go well with dwarf beans)*	Chives, leeks, garlic, onions, tomato, peppers
Broad Beans	Carrot, celery, corn, lettuce, potato	Fennel
Beets	Lettuce, onion, sage	Bean (pole and runner)
Cabbage	Beetroot, bush beans, celery, mint, onion, potato, oregano, dill, chamomile, sage. *(Aromatic plants like onion, celery, and herbs help keep cabbages pest-free.)*	Strawberry, tomato *(Although tomatoes and cabbages usually repel each other, the solanine in a few nearby tomatoes will help deter diamond-back moth larvae.)*
Carrot	Bush beans, pole beans, lettuce, onion, garlic, parsley, rosemary, peas, radish, tomato. *(Onion family plants, parsley and rosemary deter carrot rust fly.)*	Dill, parsnip

VEGETABLE	GOOD COMPANIONS	BAD COMPANIONS
Cauliflower	Peas, beans, celery, oregano. *(Peas and beans help fix nitrogen to supply to cauliflowers.)*	Nasturtium, peas, potato, and tomato
Celery	Cabbage, cauliflower, leek, onion, spinach, tomato. *(Leeks like similar high potash growing conditions as celery and celeriac.)*	Parsnip, potato
Corn	Beans, cucumber, melon, peas, pumpkin, potato, radish. *(Peas and beans supply nitrogen.)*	Tomato *(The same worm [tomato worm and corn earworm] likes both plants.)*
Cucumber	Beans, peas, celery, lettuce, pea, radish, nasturtium, corn. *(Nasturtium deters cucumber beetles and harbours beneficial spiders and beetles. Corn protects against bacterial wilt virus.)*	Cauliflower, potato, basil and any strong aromatic herbs
Eggplant	Beans, capsicum, potato, spinach, peppers	
Lettuce	Carrot, radish, strawberry, cucumber	Beans, beetroot, parsley
Onion	Broccoli, cabbage, carrot, lettuce, beets, tomato	Beans, peas

VEGETABLE	GOOD COMPANIONS	BAD COMPANIONS
Pea	Beans, carrot, corn, cucumber, radish, turnip, spinach, mint, potato	Onion family
Potato	Radish, beans, corn, cabbage, peas, eggplant	Cucumber, tomato, pumpkin, sunflower
Pumpkin	Corn, beans, peas, radish	Potato
Spinach	Celery, cauliflower, eggplant, radish. *(Leaf miners prefer radish leaves rather than spinach.)*	
Tomato	Celery, carrot, parsley, basil, marigold, garlic. *(Garlic protects against red spiders.)*	Corn, potato, fennel, and cabbage

After choosing the spot and the crop, let us also make sure we have the required tools to undertake the planting. Keeping the earth and community in mind, it is best if we can use simple tools that we can create ourselves or find locally, which are made from sustainable materials.

Thus, at the end of this step, you should have chosen and measured the site/plot and made a sketch of it to scale in your notebook. Look at the companion plant table and the sowing chart to plan which plant to sow and where. In the next chapter we will give legs to our dream and get back to earth by starting the process of planting the Garden of Hope.

READING LIST

VANDANA SHIVA, *The Violence of the Green Revolution: Third World Agriculture, Ecology and Politics*. Zed Books, 1991.

STEPHAN HARDING, *Lynn Margulis, Animate Earth: Science, Intuition, and Gaia.* Green Books, 2006.

ANNA LAPPÉ, *Diet for a Hot Planet: The Climate Crisis at the End of Your Fork and What You Can Do About It*. Bloomsbury, 2010.

ANDREA BELLAMY, *Small-Space Vegetable Gardens: Growing Great Edibles in Containers, Raised Beds, and Small Plots.* Timber Press, 2014.

Recipes for Climate Resilience

MOONG-BESAN CHILA

INGREDIENTS

1 cup split green gram
4 green chillies
2 tbsp fresh curd
1 tbsp chickpea flour/gram flour
½ tsp baking powder
A pinch of asafoetida
Salt, to taste
Oil (preferably cold pressed) for cooking

FOR THE STUFFING

4 potatoes, boiled
1 tbsp ghee
1 tbsp cumin seeds
1 onion, finely chopped
4 green chillies, chopped
½ tsp chopped coriander
½ tsp red chilli powder
1 tsp dry mango powder
Salt, to taste

METHOD

- Wash and soak the green gram overnight or for at least 3 to 4 hours.
- In a mixer-grinder, grind the soaked dal with the green chillies and curd.
- Combine the besan, baking powder, asafoetida, and salt with the ground dal and mix well.

FOR THE STUFFING

- Mash the boiled potatoes coarsely or cut them into small pieces.
- Heat the ghee in a pan over medium heat. Add the cumin seeds and onion and sauté for at least three minutes. Add the green chillies and sauté for one minute.
- Add the mashed potatoes, coriander, red chilli powder, and dry mango powder. Season to taste with salt and mix well. Remove from the heat and let the mixture cool.
- Heat a tava/griddle over medium heat. Pour a ladleful of the dal batter and spread it over the griddle.
- Put a little stuffing on each round and smear a little oil on the sides. Fold like a dosa and cook on both sides. Repeat with the remaining batter and stuffing.

AAM PANNA

INGREDIENTS

1 kg unripe mango
1½ tsp cumin powder
2 bunches coarsely chopped fresh mint
425 gm sugar
4 tsp black salt
1 litre water

METHOD

- Pressure cook or roast the mangoes and squeeze out the pulp in a big container.
- Add the cumin powder and mint leaves to the pulp and mix. Strain the mixture by squeezing it through a muslin cloth to collect the concentrate.
- Add the sugar and black salt and store the concentrate in a clean glass bottle.
- Take 2–3 tablespoons of the concentrate in a glass, add chilled water and ice cubes.

JHANGORA MILLET KHEER

INGREDIENTS

2 litres milk
500 gm jhangora/barnyard millet/samak, washed and soaked for 3 hours
200 gm sugar
1–2 drops of kewra essence
50 gm cashew, chopped
100 gm chironji/*Buchanania lanzan* (can be substituted with almonds)

METHOD

- In a heavy-bottomed pan over medium heat, add the milk and bring it to a boil.
- Drain the soaked millets and add to the milk. Reduce the heat to low and cook well, stirring continuously to avoid lumps.
- Add the sugar and cook until it is fully dissolved.
- Stir in the kewra essence to give it a sweet flavour.
- Garnish with chopped dry fruits and serve hot or warm.

Source: Pulse of Life, *Navdanya*

3

WHAT YOU SOW *is* WHAT YOU REAP

Soil, *not* Oil

'To forget how to dig the earth and tend the soil is to forget ourselves.'
MAHATMA GANDHI

Slow living begins with a return to the soil, as our identity, as the ecological basis of our being, as a source of the food that nourishes our body and provides the energy to give back to the soil, to the Earth. The harvest of giving back to the soil is our deeper sense of self as an interbeing, as connected to other beings through the soil. The harvest is well-being. The harvest is hope in our future.

When we care for the soil, we reclaim our humanity. Our future is deeply connected with the future of the Earth. It is no accident that the word 'human' has its roots in 'humus' – soil in Latin. And Adam, the first human in Abrahamic traditions, is derived from *adamah*, soil in Hebrew.

Our rupture from the soil in our minds and our distancing from the soil in our lives have led to multiple crises that threaten our well-being and our future.

In taking care of the soil, we sow the seeds of the future. As the ancient Vedas teach us, 'In this handful of soil is your future. Take care of it. It will take care of you. Destroy it. It will destroy you.'

The land is people's identity; it is the ground of culture and economy. A bond with the land is a bond with the earth – a bond that is shared by all cultures of the world. Land, soil, and food are inextricably bound together. Splitting land from soil and soil from food, treating them as separate entities was the result of a colonial idea that has led to linear economies and industrialization of agriculture.

Between 1770 and 1830, more than 3,000 bills were passed by the British Parliament to enact laws for the enclosures of the Commons. As a result of this, 6 million acres of commonly held lands, open fields, meadows, wetlands, forests, and unoccupied 'wastelands' were put into private hands and subsequently hedged, fenced, and farmed. However, enclosures are not just a historical episode that occurred in England. Enclosures have been central to the historical process of colonization. Colonialism created private property by enclosing the Commons, thus displacing and uprooting the original peoples in the Americas, Asia, Africa, and Australia.

Globalization as recolonization is leading once again to a massive land grab in various regions of the world – in India, in Africa, and in Latin America. While the dominant paradigm reduces the growing conflicts in the world to ethnic and religious violence, in addition to land grabbing, land degradation is very often at the root of these conflicts. Uprooted people are vulnerable to other exploitations, and the soils they once knew and defended are open for appropriation, despoliation, and general harm.

Industrial civilization has distracted us from considering our relation to the soil based on the arrogant belief that the more we conquer and destroy nature, the more 'developed' we are. This has led to the severing of our relationship with the land.

In contrast, indigenous cultures of Latin America have referred to land as earth and soil – Pachamama, or 'Mother Earth'. Among Dalit communities in southern India, they find their heritage in Mata Dharti (Mother Soil). Earth defenders worldwide see land as sacred because it is a living system that harbours human and animal life. For indigenous peoples, heritage is a bundle of relationships rather than a bundle of economic rights.

Living soil is a complex food web, teeming with earthworms, bacteria, and fungi. A Danish study analyzed a cubic metre of natural soil and found 5,000 small earthworms, 50,000 insects and mites, and 12 million roundworms. A gram of the soil contained 30,000 protozoa, 50,000 algae, 400,000 fungi and billions of individual bacteria. It is this amazing biodiversity that maintains, rejuvenates, and strengthens soil fertility and supports agriculture.

The living soil was forgotten for an entire century with very high costs to nature and society. A hundred years ago, the Haber plant at Ludwigshafen in Germany, run by the chemical giant BASF, transformed – by churning out ammonia in industrial quantities for the first time, triggering a green revolution. 'Bread from air' was the slogan after the discovery of the Haber-Bosch process by German chemist Fritz Haber. This led to the promotion of the illusion that the living soil was an 'empty container' for pouring synthetic fertilizers into, which were falsely seen as the source of soil fertility.

Ecological and regenerative agriculture is based on recycling organic matter, and hence, recycling nutrients. It is based on the Law of Return – giving nutrients back to the soil. As Albert Howard wrote in *The Soil and Health*: 'Taking without giving is a robbery of the soil and a banditry; a particularly mean form of banditry, because it involves the robbing of future generations which are not here to defend themselves.'

The United Nations dedicated 2015 as the International Year of Soils 'to make people aware of the crucial role soil plays in food security, climate change adaptation and mitigation, essential ecosystem services, poverty alleviation and sustainable development.' It pressed upon the global citizens and governments that we need a new pact with the earth and the soil. A pact that recognizes that we are the soil, we grow from the soil, and we are sustained by the soil. This is the new Renaissance – a new awakening that soil is alive and that taking care of it is of fundamental global importance and the most important work that humanity can do.

In taking care of the soil, we also produce more food on less land. Fertile soils are the sustainable answer to food and nutrition security. Organic agriculture is the only real answer to climate change. In the process, it also addresses food insecurity and hunger, reverses desertification, creates livelihood security by creating ecological security, and therefore, creates the path to peace. Above all, it allows a transition from the violent paradigm, structures and systems of capitalist patriarchy to the non-violent paradigm, structures and systems based on *ahimsa*, which include the well-being of all.

Just as humus in soil binds soil particles and prevents its erosion, it also binds the society and prevents violence and social disintegration. Since humus provides food, livelihood, water, and climate security, it also contributes to peace. Just as wet straw cannot be put on fire by a matchstick, communities that are secure cannot be put on fire by violent elements feeding on insecurity created by an economic model that is killing swadeshi, and is only designed for global economic powers to extract what they want.

Soil, Not Oil

'Soil is like a song. It embodies love and life. Like a song, it has the ability to move you, to pass energy reciprocally from herself to humans, animals, plants, embracing relationships and engagement from the Solar entities... when we have good soil, we have a good foundation for good health.'
FAITH KEEPER, CAYUGA NATION
Six Nations Grand River Territory, Canada

Both ecological science and our ancient wisdom teach us that all life depends on soil. However, we are under the illusion that human progress is based on how fast we can destroy, bury, and consume the soil.

Uncontrolled urbanization, mega mines, superhighways and gigantic infrastructure projects are the burial grounds of fertile soil. We are forgetting that life grows from **soul/soil**, not concrete and tarmac. Degraded and dead soil, that is soil without organic matter, soil organisms, with no water-holding capacity, creates famines and a food crisis. Thus soil, and not oil, holds the future for humanity. The oil-based fossil fuel-intensive and chemical-intensive industrial agriculture have unleashed processes that are killing the soil, and hence, impacting our future.

OIL-BASED SOCIETY	SOIL-BASED SOCIETY
Fuelled by industrial agriculture, destroying living soil through monoculture and chemicals.	Sustained by small-scale biodiverse farms and organic farming.
Law of Exploitation – of taking without giving back, creating an unprecedented ecological crisis.	Law of Return – of giving back – has ensured that societies create and maintain fertile soil, and can be supported by living soil over thousands of years.
Pestering Life – looking at every being other than human as a pest to be exterminated, poisoning the soil and air to do so, and thereby poisoning ourselves.	Bee Kind – nurturing soil; thus living in kindness with all of earth's beings: the birds, the bees, insects, worms, and microbes essential for our survival.
Supermarkets of Despair – eroding our local wisdom and connection to soil, buying food and clothes from mass retail stores.	Gardens of Hope – cultivating our own food as a declaration of hope and celebration of life, using local seeds, and participating in nurturing life.

The Living Soil: Building a Just and Peaceful Future

'Although the surface of our planet is two-thirds water, we call it the Earth. We say we are earthlings, not waterlings. Our blood is closer to seawater than our bones to soil, but that's no matter.... From the time that water invented us, we began to seek out dirt. The further we separate ourselves from the dirt, the further we separate ourselves from ourselves. Alienation is a disease of the unsoiled.'
TOM ROBBINS
Another Roadside Attraction

It has been said that if you want to destroy a people, all you need to do is destroy their soil. When people are dislocated from their territories, the intricate web that links peoples and communities to the soil is broken and can be irretrievably lost. According to the United Nations Convention to Combat Desertification (UNCCD), 40 percent of the intrastate conflicts over a sixty-year period were associated with land and natural resources. There are strong links between many of the intrastate conflicts we see today and the control and allocation of natural resources by states. According to its report, 'Desertification – The Invisible Frontline', 2014: 'An estimated 42 percent of households intensify their seasonal

mobility in the event of poor harvests, while 17 percent migrate when there is crop destruction and 13 percent leave in the case of strong climatic events such as extreme droughts.' It is believed that by 2050, 200 million people may be permanently displaced environmental migrants.

Today, it is estimated that at least half the world's population lives in cities. The urbanization process and the consequent urban expansion seem uncontainable. The agro-industrial model goes hand in hand with uncontrolled urbanization, overbuilding, urban corruption, and financial speculation; it is expelling rural populations, relegating people to slums, driving citizens away from the old towns (which become gentrified neighbourhoods), and pushing them into suburbs. Like agro-industry, the centralizing metropolis is a big sink of energy, more and more increasing land consumption per capita. Established as the dominant way of life, living and working, the metropolitan model does not offer an adequate or suitable living environment, especially in the face of climate change. It is thus important that rural communities are empowered by providing good healthcare, education opportunities, and meaningful and sustainable livelihoods so that migration to urban areas becomes unnecessary. When the importance of farmers as providers of health and builders of soil is recognized, traditional agriculture will no longer be seen as a backward and primitive activity to be conquered by industrialization and urbanization but will be given the priority it merits. Respect, honour and dignity restored to farming communities based on the new pact with the soil will create a balance between the city and the countryside.

CLIMATE CHANGE, LAND DEGRADATION, AND REFUGEE CRISIS

Before the Syrian uprising of 2011, 60 percent of Syria's land experienced the most severe and prolonged drought, causing crop failures in the land where agriculture began in the Fertile Crescent some 12,000 years ago. Extreme events like intense droughts are one aspect of global climate change. The impact of the drought was aggravated by non-sustainable use of land and water through promotion of non-sustainable agriculture. More than 80 percent crops failed and more than 75 percent livestock died, wiping out livelihoods and forcing a mass migration of more than a million farmers and herders to cities, contributing to social instability and the country's civil war. This has created 6.5 million refugees internally and 3 million refugees who have moved to neighbouring countries. In Nigeria, Boko Haram is presented as an extremist religious movement. However, as Luc Gnacadja, the head of the UNCCD, has attested, 'the depletion of Lake Chad helped create the conditions for conflict. In much of northern Nigeria, Muslim herders are in competition with Christian farmers for dwindling water supplies. The so-called religious fight is actually about access to vital resources. It is not just about Boko Haram, but in the Sahel belt you will see it is almost the same challenge in Mali and in Sudan. Furthermore, men who were or would have been gainfully employed as farmers, fishermen, fish sellers, and pastoralists have now been conscripted into Boko Haram with many of them participating in the deadly night raids of the terrorist group. Without a minimum of security of access to the land, restoration of land through investment is not possible. Peace is a prerequisite.'

TERRA VIVA, Our Soil, Our Commons, Our Future: A New Vision for Planetary Citizenship (Navdanya)

Climate Change and Kitchen Waste Composting

'I find that a real gardener is not a man who cultivates flowers; he is a man who cultivates the soil. He is a creature who digs himself into the earth and leaves the sight of what is on it to us gaping good-for-nothings. He lives buried in the ground. He builds his monument in a heap of compost. If he came into the Garden of Eden, he would sniff excitedly and say: "Good Lord, what humus!"'

KAREL APEK

The Gardener's Year

The best fertilizer is most certainly cow dung. Over centuries, farmers have used it to improve their soil fertility. However, in a city, it is not always an easy task to find this natural resource. Compost with dry leaves, stalks, kitchen waste, and any green garden residue is accessible to everyone and thus a substitute for animal manure. Before we enter into how to compost for your Garden of Hope, let us look at why you should compost your food waste with or without a garden.

What is the relationship between our food waste and climate change?

According to WWF, 'Today, an estimated one-third of all the food produced in the world goes to waste. That's equal to about 1.3 billion tons of fruits, vegetables, meat, dairy, seafood, and grains that either never leave the farm, get lost or spoilt during distribution, or are thrown away in hotels, grocery stores, restaurants, schools, or home kitchens. It could be enough calories to feed every undernourished person on the planet.' It is worth considering, then, that every time you throw leftovers away, you're not just binning tomorrow's lunch – each forkful of food was responsible for greenhouse gas emissions before it even got to your plate. Growing, processing, packaging, and transporting the food we eat together contributes to climate change. And when we throw it away, as it rots, it releases yet more greenhouse gases into the atmosphere. According to the Food and Agriculture Organization (FAO) of the United Nations, a third of greenhouse emissions globally come from agriculture, and 30 percent of the food we produce is wasted – that comes to about 1.8 billion tonnes of it in a year. If we, as a planet, stopped wasting food altogether, we would eliminate 8 percent of our total emissions. By simply rethinking how you shop and cook, it should be possible to reduce the amount we waste and so lessen our contribution to climate change.

How you throw your food away also matters. Organic matter rotting in a landfill releases methane, a greenhouse gas several times more potent than carbon dioxide. But if you compost your leftovers in a well-maintained bin that lets in oxygen, you will significantly reduce the amount of methane released into the atmosphere, and the carbon in the composting organic matter will be held in the resulting soil.

FIGHT CLIMATE CHANGE & BECOME A KEEPER OF THE SOIL

To compost is to help the cycle of life. Plants mine the soil to obtain the nutrients they require for healthy growth. If what the plant takes from the soil is returned to it, the fertility of the soil can be maintained. Composting is an easy way to give back to the soil what has been taken from it during plant growth.

One study estimated that the greenhouse gas emissions from composting are just 14 percent of the same food dumped into landfills.[1] Project Drawdown, a research organization that identifies potential solutions to climate change, estimates that if composting levels increased worldwide, by 2050 we could reduce emissions by 2.1 billion tonnes.

According to the United Nations Environment Program (UNEP), food loss and waste generate an estimated 8–10 percent of global greenhouse gas emissions, while using land and water resources increasingly puts pressure on biodiversity. The low-hanging fruit in the food system thus is to move from landfill disposal of food waste to anything else that brings down the emission, which includes home composting, anaerobic digestion, or incineration. Whatever you do, anything is better than throwing it in your kitchen dustbin, from where it ultimately reaches the landfill. While there are environmental differences between those other disposal methods, the magnitude in volumetric terms is much smaller, and thus the positive impact of it is insignificant. For this switch to have any significant impact, we need individuals and governments to commit to composting at a far larger scale than present today.

Data going back to 1940, as reported by Eco Farming Daily, shows: 'The level of every nutrient in almost every kind of food has fallen between 10 and 100 percent. An individual today would need to consume twice as much meat, three times as much fruit and four to five times as many vegetables to obtain the same amount of minerals and trace elements available in those same foods in 1940.'

In humans, our 30 trillion or so human cells are outnumbered by diverse microorganisms, our 'inner soil' that function throughout our bodies in exchange for food (bacteria in our colons prefer plant fibre!) and shelter. Our microbiome performs a wide range of critical functions, such as bolstering our immune systems, manufacturing vitamins, supporting digestion, and protecting against disease.

Thus, if we eat diverse and clean (pesticide-free), we will compost well too, resulting in the waste that comes out of our kitchen becoming the food that our kitchen garden plants will eat.

Everything you put into the compost is primarily made of carbon, nitrogen, water, and oxygen in varying amounts. Carbon compounds (browns) are present in plant residues (through lignin and cellulose) and more nitrogen compounds (greens) in kitchen wastes and animal wastes (dung and urine). The ratio of browns to greens should be 20:1 or 30:1 for a rich and successful compost.

CARBON

Carbon-rich matter (like branches, stems, dried leaves, peels, bits of wood, bark dust or sawdust pellets, shredded brown paper bags, corn stalks, coffee filters, coffee grounds, conifer needles, egg shells, straw, peat moss, and wood ash) gives compost its light, fluffy body.

NITROGEN

Nitrogen or protein-rich matter (manure, food scraps, green lawn clippings, kitchen waste, and green leaves) provides raw materials for making enzymes.

To consider what your composting options are, you must consider the space at your disposal.

---

1. *Resources, Conservation and Recycling*, vol. 52, no. 12, October 2008, pp. 1373–81.

WHERE YOU LIVE	WHAT ARE YOU COMPOSTING THE MOST?		
	Composting mostly kitchen waste	Kitchen waste and some yard scraps	A lot of yard scrap
Urban, not much outdoor space	Worm bin		
Urban, with some outdoor space (balcony or patio)	Worm bin, compost tumbler or earthen pot composting	Compost tumbler	
Suburban with yard	Compost tumbler, enclosed bin	Compost tumbler, enclosed bin	DIY enclosed bin (large)

4 COMPOSTING CONCERNS AND SOLUTIONS

1. SPACE

Even if you live in an apartment or other urban location, you can reduce your organic waste and produce your own compost for your plants or your plot in the local community garden. There are several different methods for composting indoors, which are odour-free and compact enough for even smaller apartments. Vermicomposting and the three-bin system may work best for such spaces.

2. SMELL

Though the unpleasant smell from your compost pile can be a deterrent for you to take up composting, you can reduce or eliminate odour completely by following these two practices. First, remember to not put bones or meat scraps into the compost.

Second, cover new additions to the compost pile with dry grass clippings or similar mulch. Adding lime or calcium will also neutralize odours. If the compost smells like ammonia, add carbon-rich elements such as straw, peat moss, or dried leaves. A vermicomposting bin, however, ensures that the problem of foul odour is almost negligible. Also remember, the higher the nitrogen, the stronger the smell; balance the ratio and the problem is resolved.

3. FRUIT FLIES

Another deterrent to composting are fruit flies, which are naturally attracted to the compost pile. Discourage them by covering any exposed fruit or vegetable waste. Keep a small pile of grass clippings next to your compost bin, and when you add new kitchen waste to the pile, always cover it with one or two inches of clippings. Adding lime or calcium will also discourage flies.

4. ATTRACTING RATS

Food waste that invites varmints includes meat, poultry, fish, fat, oil, dairy products, bread, grains, and bones. If this is a problem for you, avoid composting food that attracts unwanted critters. The more actively you manage and turn your pile during early decomposition, the less likely you will have problems.

As a first-time composter, you cannot start by trying to compost everything. Do not compost meat, bones, or fish scraps unless you are using a composter designed specifically for this purpose, otherwise, you will be disappointed, as they take very long to compost besides becoming a source of pests and rodents. Start simple, with kitchen waste from vegetable peels, and slowly graduate into more complex composting. As composting is really about patience, you don't want to test yours to the limit right in the beginning.

Connecting with Soil: Preparing the Garden of Hope

'The first supermarket supposedly appeared on the American landscape in 1946. That is not very long ago. Until then, where was all the food? Dear folks, the food was in homes, gardens, local fields and forests. It was near kitchens, near tables, near bedsides. It was in the pantry, the cellar, the backyard.'

JOEL SALATIN

American author and farmer

After carefully planning your garden on paper, it is now time to get your hands dirty. Gardening is, after all, about tending to the earth and soil so that you can witness the miracle of life in your backyard.

The preparation of the soil may be the most important step of your garden experience to guarantee good crops. If you want to improve your soil and make it more suitable for your vegetable garden, you first have to figure out what you are dealing with. Is your soil sandy or clay-based? Is it too acidic or too alkaline? You can learn this through testing your soil.

SOIL TESTING

Soils are classified as clay soils, sandy soils, or loamy soils.

Clay is nutrient-rich, but slow-draining.
Sand is quick draining but has trouble retaining nutrients and moisture.
Loam is generally considered to be the ideal soil because it retains moisture and nutrients but doesn't stay soggy.

To determine your soil type, take a handful of moist (but not wet) soil from your garden and give it a firm squeeze. Then, open your hand. One of three things will happen:

• It holds its shape, and when you give it a light poke, it crumbles. Lucky you – this means you have luxurious loam!

• It holds its shape and, when poked, sits stubbornly in your hand. This means you have clay soil.

• It falls apart as soon as you open your hand. This is sandy soil.

• Scoop some soil into a container. Then add half a cup of vinegar. If the soil bubbles or fizzes, it's alkaline.

NOW THAT YOU KNOW WHAT TYPE OF SOIL YOU HAVE, YOU CAN WORK ON IMPROVING IT.

Good soil is 50 percent solids and 50 percent porous space, which provides room for water, air, and plant roots. The solids are inorganic matter (fine rock particles) and organic matter (decaying plant matter). The inorganic part of the soil can be divided into three categories based on the size of the particles it contains. Clay has the smallest soil particles; silt has medium-sized particles; and sand has the coarsest particles. The amount of clay, silt, and sand in a soil determines its texture. Loam, the ideal garden soil, is a mixture of 20 percent clay, 40 percent silt, and 40 percent sand.

For harvesting a bigger and better crop of vegetables, you could improve the texture and structure of your soil. This improvement, whether to make the soil drain better or hold more water, is best achieved by adding organic matter to the soil.

Fertilizers strip soils of their natural fertility. Inorganic fertilizers, like nitrogen-based fertilizers, kill soil fauna that recycle nutrients naturally. Fertilizers also cause degradation in the form of nutrient depletion, salt accumulation, and acidification; thus, to ensure healthy and organic soil, we must learn to compost and make our own manure and natural fertilizers.

Once your soil is prepared and duly revitalized with the help of the organic compost, it is time to revisit your garden plan. Draw out the plan on the soil. If you want, you can use stakes and strings and lay out the rows.

Consider making a raised bed garden in your yard or on your patio. There are several advantages to this type of garden. Though there is more preparation and effort in the beginning, it's more rewarding!

- Raised bed gardens are easy to plant and tend.

- You can fill the raised bed with an ideal mixture of garden soil, mulch, and compost.

- You can also grow more vegetables in a smaller area, as you do not need to leave room for walking between the rows of plants.

- Raised bed gardens should be about 48 inches across so that you can reach into the centre of the bed from either side.

TIP: Plant tall-growing vegetables such as corn and green beans on the north end of the garden to prevent them from shading shorter, medium-height plants such as tomatoes, cucumbers, peppers; eggplants in the centre; and shorter vegetables such as lettuce, radish, and onions on the south end of the garden. The garden layout thus can run from east to west, but the rows of vegetables should always run from north to south for maximum sun exposure.

Many vegetables, including peas, pole beans, squash, cucumbers, gourds, and melons, will naturally climb a support and grow up rather than out, leaving more ground space for other plants. Support structures include trellises, strings, and tepees made from poles, chicken wire, or a chain-link fence. Tomatoes also can be trained to grow upright in wire cages or tied to stakes.

SOWING THE SEED

In the garden or in pots, sow seeds generously to ensure germination. While sowing, scatter the seeds or plant them in furrows or hills. As a general rule, plant a seed to a depth of not more than three or four times its thickness. If planted too deep, the seeds may germinate but die before reaching the surface. If planted too shallow, wind or rain may blow or wash the seeds away before they sprout. In sandy or lighter soils, plant a little deeper. Light mulch over the newly planted row will help maintain moisture.

Plant your heirloom seeds based on your garden plan and add a layer of mulch to help retain water and prevent weeds.

Mulching is one of the most natural and easiest gardening techniques used to protect an organic vegetable garden from weeds, drying out, pests, extreme weather, and erosion.

High-quality mulch adds invaluable nourishment to the soil as it safeguards it.

Mulches help keep the soil alive by attracting earthworms and microbes, which help loosen soil structure and supply essential nutrients.

- Before mulching, weed and water your organic vegetable garden. Fertilizer should also be applied before mulching, as organic materials decompose quicker under a layer of mulch.

- Mulch should not be layered on so heavily that it blocks moisture, air circulation or sunshine from penetrating the soil. Apply smaller-grade organic mulches such as wood chips or sawdust up to three inches in thickness. Spread lighter mulches like leaf mould, straw, and hay as much as six inches in depth.

- A rain barrel is a practical addition to an organic vegetable garden to collect water for your plants.

TRANSPLANTS – BUY OR GROW THEM YOURSELF

Many crops, such as tomatoes, eggplants, peppers, and cabbage, can be started indoors and later transplanted into the garden. This head start permits the long-season crops to grow and mature better. Some people choose to grow their own transplants. This allows the gardener to select specific cultivars and control seedling growth. In addition, many gardeners get personal satisfaction from germinating and growing their own transplant.

Your garden is now ready for you to look after, water, and keep healthy as you watch the miracle of life unfold before your eyes! In three to six weeks' time, you will be tending the plants of your Garden of Hope. In the next chapter, we will explore how to tend to your garden and look after the water cycle by not polluting it or wasting water while irrigation.

4 WAYS TO TACKLE DEFORESTATION

Soil is the earth's fragile skin that anchors all life on Earth. It is made of countless species that create a dynamic and complex ecosystem and is among the most precious resources to humans. Increased demand for agriculture commodities generates incentives to convert forests and grasslands to farm fields and pastures. The transition to agriculture from natural vegetation often cannot hold onto the soil, and many of these plants, such as coffee, cotton, palm oil, soybean, and wheat, can actually increase soil erosion beyond the soil's ability to maintain itself.

Half of the topsoil on the planet has been lost in the last 150 years. In addition to erosion, soil quality is affected by other aspects of agriculture. These impacts include compaction, loss of soil structure, nutrient degradation, and soil salinity.

Every forest in the world today is affected by human activity at some level, either directly by destruction or the introduction of invasive species, or indirectly by the impacts of climate change. Forest restoration – the process of improving the health, productivity, and array of life of a forest – is a complex undertaking that can never fully bring back the original forest. That's why it's far better to conserve existing healthy forests and prevent them from being degraded or destroyed in the first place.

1. PLANTING TREES

This is the most crucial part of forest restoration, though funding for doing so often runs dry. Many reforestation programmes need to plant seedlings and grow their own native trees. So, get involved and adopt your city forest or the forest near you; learn about its ecosystem and plant trees accordingly. As many as you can, whenever you can. Look after the trees that are being planted. As Rabindranath Tagore said, 'The one who plants trees, knowing that he will never sit in their shade, has at least started to understand the meaning of life.'

2. IMPROVING SOILS

Soils need microbes and small bugs like centipedes, beetles, and worms to thrive. Adding organic matter to the soil can radically change a forest and support restored forest health.

3. PROTECTING WILDLIFE CORRIDORS

Looking at how plants and animals move and reproduce across landscapes is key to forest restoration. A corridor the width of a swimming pool can allow animals to roam among forest fragments, significantly improving their chances of surviving, reproducing, and flourishing.

4. SUSTAINABLE MANAGEMENT OF LAND

Successful restoration must address the needs of farmers who care for small plots of land and whose access to resources profoundly affects forests. Working with governments to promote practices like agroforestry – a farming technique that incorporates the cultivation and conservation of trees among crops or pasture land for more productive and sustainable land use – provides huge benefits to forest restoration efforts.

LIFE LESSONS FROM THE CHIPKO MOVEMENT

While I was doing my PhD, I used my scholarship money to come back every summer and winter to volunteer with Chipko. Growing up as a forest officer's daughter, I always felt that the forest was beauty and peace. But working with the women of Chipko, I realized that the forest is also life, livelihood, and knowledge. Everything I learned about building movements came from Chipko.

It taught me about self-organizing. Women would gather daily and put a fistful of grain in a common basket, and that would be the food they ate as they protested in the forest. If one woman was involved with the protest, others would take care of her children, cows, buffaloes; and then they would rotate. I learned how you have to turn within yourself to see what resources you have, rather than looking externally. Chipko taught me the value of self-help and solidarity. I remember in 1977 this forest in Haldwani, Uttarakhand, was going to be logged for timber. And because the government had already faced some defeat, they decided to hire local contractors from within the village, rather than from outside. Timber was still being logged, and the village headman was given the contract for logging. Yet his wife, Bachni Devi, led the protest to stop the logging. The women arrived en masse, carrying lanterns. Seeing them, the police present said, '*Foolish women, can't you see the sun is out?*' '*These lanterns are not for the sun,*' said the women. '*They're for you.*' The slogan they used was '*Kya hain jungle ke upkaar? Mitti, paani, aur bayaar.*' (What are the jungles' gifts? Soil, water, and pure air.) In other words, forests are not resin, timber, and revenue. Chipko taught me ecology and biodiversity. When we finally got logging banned after the 1978 floods, it was because the authorities finally understood what the women were saying: Forests regulate water, and therefore, to avoid floods and droughts, we need to protect them. While others had treated the forest as separate from the river, the women knew it was connected. Forty years later, this is the language of science. Today people look at the ecological functions of the forest. Scientists took half a century to catch up with illiterate women. And it is primarily because people have a tendency to essentialise, or assign a particular quality as intrinsic to a person. They say, '*You haven't been to school, you're no good.*' Or, '*You're a woman in a village? You're a nobody.*' But the Chipko women were shaping the future of India. It was their work that gave us the Ministry of Environment and Forests and the concept that forests are for conservation, not extraction. Finally, it taught me humility. There is a certain arrogance that comes with science. Science is always pitted against mythology, but it is the mythology of this time. I was a part of the arrogance of science when I was working with nuclear reactors. But Chipko taught me that everyone has knowledge and that the role of science is merely to discover that knowledge.

Source: https://feminisminindia.com/2019/10/15/vandana-shiva-interview-chipko-movement/

WATCH LIST

JHING CHIK JHING by NITIN NANDAN, a 2010 Marathi film about a boy who grows up amidst farmers' suicides and debt in Vidarbha.

DO BIGHA ZAMIN by BIMAL ROY, a Hindi classic released in 1953.

NIYAMGIRI: THE MOUNTAIN OF LAW by SURYA DASH, a ten-minute short film released in 2018.

Recipes to Honour the Living Soil

GOODNESS OF MUSTARD SEEDS *in* BENGALI-STYLE EGGPLANT

INGREDIENTS

1 large eggplant/ aubergine
1½ tsp black mustard seeds, plus 1 cup water
¼ tsp red chilli powder
4–5 tsp mustard oil
1 tsp panch phoron masala
1 cup curd/yoghurt
1½ tsp salt
Sprinkle of black pepper
Green cardamom powder (optional)

METHOD

- Cut the eggplant into 1 inch cubes.
- Grind the mustard seeds in a coffee grinder and soak the ground mixture and red chilli powder in a cup of water.
- Heat the mustard oil in a pan over medium heat. Add the panch phoron masala and after a few seconds, add the prepared black mustard and chilli water. This will splutter, so cover the pan with a lid.
- Now add the eggplant and stir in more water if required. Cover the pan with a lid and cook for ten minutes.
- When the eggplant is cooked, add a cup of yoghurt and season to taste with salt.
- Stir the mixture well to combine and heat up the yoghurt, but do not boil. Garnish with a sprinkle of black pepper and green cardamom powder. This dish also tastes good cold the next day.

Locally pressed oils like mustard and sesame are a good switch to help protect the planet and stop desertification of important ecosystems.

Source: Tilhan (Oilseeds), *Navdanya*

RESTORE LAND WITH THE *pulse of* VEGETARIAN **DHANSAK**

INGREDIENTS

¼ cup split pigeon peas
¼ cup split green gram
¼ cup white gram
¼ cup red lentil
½ cup chopped broccoli
½ cup chopped carrots
½ cup chopped celery
1 tbsp oil
¼ tsp cumin seeds
1 tbsp chopped green chillies
½ cup chopped onions
1 tbsp ginger-garlic paste
½ cup chopped tomatoes
¾ cup diced green pepper
½ cup diced eggplant/ aubergine
1 tbsp garam masala
Fresh coriander, for garnish
Salt, to taste
½ tsp ghee (optional)

This vegetarian version of the dish encourages restoration of the ecology by adding more pulses to our daily diet.

METHOD

- Wash all the lentils together.
- In a pressure cooker over medium heat, add 2 to 2½ cups of water with the drained lentils. Add the broccoli, carrots, and celery. Close the lid of the cooker and cook for 2 whistles.
- Heat the oil in a deep pan over medium heat. Add the cumin seeds, green chillies, and onion; fry until the onions become golden brown.
- Now add the ginger-garlic paste and cook for some more time.
- Add the tomatoes, green pepper, and eggplant. Stir in some salt and cover the pan with a lid.
- Stir occasionally to check if the vegetables are done. Transfer the cooked dal from the pressure cooker.
- This is the time when you can stir in some warm water. Add the garam masala, fresh coriander, and season to taste with salt. Stir until it boils. For a great taste, you can top it with ghee.

Source: Pulse of Life, *Navdanya (2016)*

HARVEST *of* HOPE

MIXED VEGETABLE CURRY

INGREDIENTS

1 cup mixed vegetables sliced, boiled or steamed (carrot, cauliflower, beans, peas, potato, etc.)
½ cup grated coconut
3 green chillies
1 tbsp sesame seeds
1-inch piece cinnamon
2 cloves
2 tbsp butter
½ tsp mustard seeds
½ tsp grated ginger
½ tsp crushed garlic
Salt, to taste
½ tsp red chilli powder
1 cup sliced tomatoes

METHOD

- Drain the boiled vegetables and keep the stock aside.
- In a mixer-grinder, blend together the grated coconut, green chillies, sesame seeds, cinnamon, and cloves to a smooth paste.
- Heat the butter in a skillet over medium heat. Add the mustard seeds and allow them to splutter.
- Add the ginger, garlic, and the prepared coconut paste. Cook for 3–4 minutes.
- Add the boiled vegetables and mix well to combine.
- Add ½ cup stock. Cover the pan and let simmer for 5 minutes.
- Season to taste with salt and add the red chilli powder, tomatoes, and cook until the gravy is thick.

This is an easy recipe for you to enjoy from the Garden of Hope you have planted. You can substitute the vegetables with any vegetable that you have grown in your garden.

Source: Tilhan (Oilseeds), *Navdanya*

4

WATER *as* COMMONS

With the world's growing population, increasing pollution, and overuse of water, the available freshwater supply is becoming more and more limited. A 2009 publication sponsored by the World Bank's International Finance Corporation, as well as a number of multinational corporations, including Nestlé S.A. and the Coca-Cola Company, found that by 2030, global freshwater demand will exceed availability by 40 percent.[1] These big corporates are using this information to commodify water and stake claims on the precious 'commodity'. But water is not a commodity but a Commons. The Commons include any natural resources that are not owned by an individual or corporation.

Water is being privatized and sold as a commodity in many ways. When a company takes a river and uses it for hydropower generation, and steals it from the community and from the society that was nourished by that river, it is a form of water privatization. River diversion is also a form of privatization. Public water supplies are increasingly pressured to privatize their services as local freshwater sources are bought by these same companies, and global trade agreements and international development organizations increasingly promote the privatization of freshwater supplies. Also, there's the outright privatization through World Bank-influenced contracts to companies like

Bechtel to privatize urban supply. Freshwater has gone through some significant changes over the past twenty years since 1995. What once had been sold by the gallon as an emergency storm supply in grocery stores in the U.S. is now marketed by the pint by global corporations. Bottled water is privatized water. We should have access to clean wells, clean rivers, and public supply that is clean, drinkable water. That we are reaching the situation that corporations are using our public water bodies like rivers and wells to basically dump pollution, and then us being forced into depending on bottled water is a two-way privatization: first by the polluters and then by the bottlers.

Destruction of water resources and of forest catchments and aquifers is one form of terrorism. Denying poor people access to water by privatizing its distribution or polluting wells and rivers is also terrorism. Greed and appropriation of other people's share of the planet's precious resources are at the root of conflicts worldwide.

Every day, 2 million tonnes of sewage as well as industrial and agricultural waste are discharged into the world's water (UN WWAP 2003), the equivalent of the weight of the entire human population of 8.2 billion people. The UN estimates that the amount of wastewater produced annually is about 1,500 km^3, six times more water than exists in all the rivers of the world (UN WWAP 2003). Unsafe water causes 4 billion people to suffer from diarrhoea each year, and results in 2.2 million deaths, mostly of children under five. This means that 15 percent of child deaths each year are attributable to diarrhoea – a child dying every fifteen seconds. In India alone, the single-largest cause of ill health and death among children is diarrhoea, which kills nearly half a million children each year (WHO and UNICEF, 2000). Thus, we see global corporate greed coupled with loss of traditional systems of community share; hence, community care for water bodies is declining where water is no longer a part of the Commons but a commodity to be profited from. This leads to grave inequality between those who can and those who cannot afford clean drinking water, let alone the lack of understanding that water belongs to all creatures and cannot be monopolized by human beings.

Water works as a democracy. Rain doesn't just fall on Washington; it distributes itself. Climate change is making it a little chaotic, but water comes to us in a democratic way, whether it's through rivers or precipitation. That's the way we need to look at water. It is one of the most important Commons in the world and embodies the kind of democracy and equality that we want to see in the world and what we are creating for the next generation. We need to recognize that the best place to protect water is at the level of local community action, working in cooperation and solidarity with the community and for the community, and then working upwards. And that's how water works. Rivers should be able to flow; water should be able to flow through the sap of plants. After all, 70 percent of every species is water; that 70 percent of water entitlement for all beings needs to be secured. It can't be secured if rivers are eaten up. It can't be secured if water is privatized and only those people who can afford to pay for it have access to it and the other species, the other human beings just wither away.

---

1. The 2030 Water Resources Group. "Charting our Water Future: Economic Frameworks to Inform Decision-making," 2009.

Marine Pollution
5 Problems and 5 Solutions

'The last fallen mahogany would lie perceptibly on the landscape, and the last black rhino would be obvious in its loneliness, but a marine species may disappear beneath the waves unobserved and the sea would seem to roll on the same as always.'
G. CARLETON RAY
'Biodiversity', National Academy Press (1988)

Plankton are not just one species of sea creature but, rather, a large variety of tiny organisms. Algae, bacteria, crustaceans, molluscs and more are all considered plankton. What most of us may be unaware of is that phytoplankton grow and get their energy through photosynthesis and are responsible for producing an estimated 80 percent of the world's oxygen. Thus, oceanic pollution directly impacts the quality of air on earth.

Marine pollution is a growing problem in the world we inhabit. Our oceans are being flooded with two main types of pollution: chemicals and trash. Chemical contamination, or nutrient pollution, is concerning for health, environmental, and economic reasons. It occurs when human activities, notably the use of fertilizer on farms, lead to the runoff of chemicals into waterways that ultimately flow into the ocean. Marine trash encompasses all manufactured products – most of them plastic – that end up in the ocean. Littering, storm winds, and poor waste management all contribute to the accumulation of this debris, 80 percent of which comes from sources on land. The most common types of marine debris include plastic items like shopping bags and beverage bottles, along with cigarette butts, bottle caps, plastic straws, toothbrushes, food wrappers, and fishing gear. Plastic waste is particularly problematic as a pollutant because it takes hundreds of years to decompose. There's so much junk at sea; the debris has formed giant garbage patches. There are five of them around the world, and the largest – the Great Pacific Garbage Patch – includes an estimated 1.8 trillion pieces of trash and covers an area four times the size of the state of Rajasthan.

Noise pollution generated by shipping and military activity can cause cellular damage to a class of invertebrates that includes jellyfish and anemones. These animals are a vital food source for tuna, sharks, sea turtles, and other creatures.

So, now let's look at five major culprits of marine pollution, the role we play in the spread of this menace, and what we can do to curb it:

1. MORE PLASTIC THAN FISH

Eight million metric tonnes of plastic is dumped into the oceans each year. That's about 17.6 billion pounds – or the equivalent of nearly 57,000 blue whales – every single year. By 2050, ocean plastic will outweigh all of the ocean's fish. This trash poses dangers to both humans and animals. Fish become tangled and injured in the debris, and some animals mistake items like plastic bags for food and eat them. Small organisms feed on tiny bits of broken-down plastic called micro-plastic, and absorb the chemicals from the plastic into their tissues. When larger animals eat small organisms that consume micro-plastics, the toxic chemicals become part of their tissues, making the micro-plastic pollution migrate up the food chain and eventually becoming part of the food that humans eat.

Solution: Reduce, Refuse, Recycle Single-Use Plastics.

i. Recycling helps keep plastics out of the ocean and reduces the amount of 'new' plastic in circulation. At present, just 9 percent of plastic is recycled worldwide. This should go without saying, but when you use single-use (and other) plastics that can be recycled, always be sure to recycle them.

ii. Reduce the use of plastic wherever possible by carrying with you

- Grocery bags and dry-cleaning garment bags
- Reusable bottles
- Reusable utensils for take-away food, coffee cups
- Reusable straws and cutlery (e.g., stainless steel straws).

iii. Stay informed on issues related to plastic pollution and help make others aware of the problem. Tell your friends and family about how they can be a part of the solution, or host a viewing party for plastic pollution-focussed documentaries, like *Bag It, Addicted to Plastic, Plasticized*, or *Garbage Island.*

2. OIL RUNOFF FROM VEHICLES

Road runoff oil from our vehicles is carried to the sea by drainpipes and rivers, and these are three times more polluting than the headline-grabbing oil spills which account for just 12 percent of the oil in our oceans. Most cars drip oil onto the ground, usually on impermeable concrete or asphalt, and that oil ends up trickling into the ocean. In drier places, the oil builds up on the asphalt and when it finally rains the water shuttles large amounts of oil into the ocean. Improper disposal of used oil, which includes oil leaking from cars, contributes significantly to storm-water pollution.

Solution: Check your runoff quotient and minimize it.

i. If you have an oil leak, do your part and have it repaired.

ii. Handle and discard your motor oil and other automotive

fluids responsibly. The simplest way is to give it to your nearest workshop and oil dealership for recycling.

3. POLLUTION IS IN FASHION (LITERALLY)

With each load of laundry, more than 700,000 synthetic microfibres are washed into our waterways. Unlike natural materials such as cotton or wool, these plasticized fibres do not break down. One study shows that synthetic microfibres make up as much as 85 percent of all beach trash.

Solution: Wear natural fibres.

4. EVEN NUTRIENTS CAN BECOME HARMFUL

When dumped at sea in large quantity, agricultural nutrients from chemical farming, such as nitrogen, can stimulate the explosive growth of algae. When algae decompose, oxygen in the surrounding waters is consumed, creating a vast dead zone that can result in mass die-offs of fish and other marine life.

Solution: Eat and grow organic and pesticide-free food.

5. GARDEN POLLUTION

Careless or unnecessary use of lawn and garden fertilizers can contribute to nitrogen and phosphorus pollution of streams, rivers, lakes, estuaries, and coastal waters.

Solution: Eliminate chemical fertilizers and pesticides from your lawn or garden.

Also follow these simple guidelines:

i. Don't overwater your lawn, as that can create extra runoff as well.
ii. Try to incorporate permeable pavement (which allows runoff through it into the soils below), as well as rain barrels to collect water, and rain gardens, grassy swales and driveway-side vegetative strips – all planted with region-appropriate native plants, of course – to help filter contaminants out before they hit the storm sewers.

3 Ways to Protect Water as a Commons

'Water is the driver of nature.'
LEONARDO DA VINCI

1. SAY NO TO BOTTLED WATER

Thc right to water has been recognized in international law. Yet, some companies do not respect this. The chairman of the board of Nestlé, for example, once said, 'The one opinion, which I think is extreme, is represented by non-governmental organizations (NGOs), who bang on about declaring water a public right. That means as a human being you should have a right to water. That's an extreme solution.' By buying bottled water you support companies such as Nestlé who steal water from the public and let you pay a high price for it. Only a generation ago water was available for free. We, however, already pay INR 20–25 for one litre of bottled drinking water. One wonders how much the next generation will have to pay.

According to the Food and Drug Administration of India, 53 percent of packaged drinking water in Mumbai was found to be unfit for human consumption. In many ways, bottled water is only as regulated as tap water in India. Even in cases where regulations exist, there are serious loopholes in implementation. In many cases, water churned out from a single source gets to be sold under different brand names. To think one brand is better than another is delusional. Most of the bottled water sold in India hardly contains any minerals. In the process of reverse osmosis (RO), most of the minerals are lost. This is the reason most bottled water bottles have 'packaged drinking water' and not 'mineral water' written on them. This water has no intrinsic value except an assurance of safety, which we know is a myth. Thanks to advertising and celebrity endorsements, we are made to believe that all water except bottled water is unsafe.

Worldwide, some 2.7 million tonnes of plastic is used to bottle water each year. Even in the US, six out of seven plastic bottles consumed are 'downcycled', and not recycled, meaning they end up in garbage dumps, landfills, or even worse, in the oceans. Many Indian villages have complained about depleting groundwater levels after bottled water companies began extracting water for sale in neighbourhood areas. Even in areas with proper recycling facilities in place, millions of plastic bottles end up polluting the environment. These bottles eventually enter our waterways where they break down into smaller particles but never degrade as, for example, an apple or a newspaper would. The bottles break into ever smaller particles that marine animals usually mistake for food.

It takes about 2.5 decilitres of oil to manufacture one plastic bottle of one litre capacity. This oil comes from deep down in the Earth. So first of all, you need to drill a well. Not only does this process contribute to CO_2 emissions, it also disturbs the local environment. Oil must be pumped up, then the raw oil must be processed to eventually end up as one of the raw materials for producing that water bottle, which we buy and then throw away as soon as it's empty. Isn't it crazy how we prefer to put so much effort in manufacturing and shipping that bottle while in a lot of places you can just walk to the tap to fill your glass or bottle?

Many communities have had no option but to go to court to try and protect their groundwater from corporate water bottlers.

Extremely expensive and time-consuming, these legal battles have torn towns apart. Although some communities have banned commercial water extraction, not all towns have had such success.

According to Euromonitor International, Nestlé is globally the biggest water bottler in terms of sales, followed by Coca-Cola, Danone, and PepsiCo. Nestlé Waters, the Paris-based subsidiary, owns almost fifty brands, including Perrier, S. Pellegrino, and Poland Spring. The company, it seems, has been preparing for shortages for decades. Helmut Maucher, the former CEO of the multinational, in an interview with the *New York Times*, said, 'Springs are like petroleum. You can always build a chocolate factory. But springs you have or you don't have.' This blatant monopoly of water can be stopped if we drastically cut our use and choose to purify our water at home and carry our own water bottle when we travel.

2. COLA-FREE ZONES

Remember as you grab a can of aerated cola or soda from well-known international brands that it takes almost three litres of water to make one litre of the beverage in question. In order to satisfy this need, the companies are increasingly taking over control of aquifers in communities around the world, robbing the fields, the ecosystem, and people of their water. Sugar, which is used in excess in these products, comes from water-guzzling crops that further increase the total amount when we calculate the amount of water needed for the production of these drinks. According to our research, Coca-Cola is the number one buyer of sugarcane in India, and Pepsi is number three. If you take into account the water used for sugarcane plantation, then we are using almost 400 litres of water to make a bottle of Coca-Cola.

In 2008, Coca-Cola conducted an assessment of a factory and found it used 35 litres of water to make every half-litre of its beverage. The researcher who worked on that assessment says the company pressured him to adopt a 'net green' accounting method that would have lowered the water footprint of its agricultural supply chain. Coming under pressure from local communities and environmentalists, Coca-Cola pledged to become 'a truly water-sustainable business on a global scale' in 2007. Eleven years later, by the standards of the 2008 assessment, the company still has nearly 99 percent of its water footprint to go.

PepsiCo and Coke have been accused of dehydrating communities in their pursuit of water resources to feed their own plants, drying up farmers' wells, and destroying local agriculture. The companies have also violated workers' rights in countries such as Colombia, Turkey, Guatemala, India, and Russia. Only through their multi-million dollar marketing campaigns can the cola companies sustain the clean image they crave.

3. WATER HARVESTING AND CONSERVATION – PROTECTING THE COMMONS

The 'Charting Our Water Future' report by global consultancy McKinsey predicted that by '2030, under an average economic growth scenario and if no efficiency gains are assumed, global water requirements would grow from 4,500 billion m^3 today (or 4.5 thousand cubic kilometres) to 6,900 billion m^3.' As the report states, 'the drivers of this resource challenge are fundamentally tied to economic growth and development. Agriculture accounts for approximately 3,100 billion m^3, or 71 percent of global water withdrawals today, and without efficiency gains will increase to 4,500 billion m^3 by 2030 (a slight decline to 65 percent of global water withdrawals). The water challenge is therefore closely tied to food provision and trade.' While being mindful of eating seasonal fresh food which does not rely on water-intensive chemical farming is definitely the call of the hour, proactive measures for water conservation need to also be taken both at an individual and collective level.

THE CASE OF PLACHIMADA AND COCA-COLA COMPANY

In June 2019 a news report conducted by 24/7 Wall St. implied that Coca-Cola and PepsiCo top the list of largest ocean plastic polluters in the world. This report states that Coco-Cola is the top producer of ocean plastic using 3 million tonnes of plastic packaging annually, with PepsiCo taking second place. Nestlé, another food brand, also makes it to the list of 20 corporate offenders in ocean water pollution. This statistic may still feel far removed from our own lives so let us look at the Plachimada Movement to understand how it impacts us. In March 2000, Coca-Cola, under its Indian subsidiary Hindustan Coca-Cola Beverages Private Limited (HCCBPL), started operations at its bottling plant at Plachimada in the southern state of Kerala. Over the next few years, the area surrounding the plant began to feel the hazardous effects, as groundwater was contaminated and toxic waste released. The bottle washing at the plant involved chemicals, and the resulting sludge was taken out of the plant. Initially, the sludge was sold as fertilizer to unsuspecting farmers, following which it was given free, and with increasing resentment among villagers, it was merely dumped on the roadside. In the few months since commencement, more than 1,000 families in the surrounding villages had been affected with diseases linked to toxicity. In 2003, the villagers organized themselves and took the company to court. In a few months, the Perumatty panchayat revoked the plant's license. Coca-Cola took the case to the Kerala High Court, which at first sided with the firm, saying the panchayat's claims were unscientific and unfounded. The legal battle lasted years. While these legal battles were going on, a World Water Conference was organized near Plachimada at Pudussery on 23 January 2004. On the third day of the conference, the Plachimada declaration was adopted, and it made several significant claims: 'It is our fundamental obligation to prevent water scarcity and pollution and to preserve it for generations.... Water is not a commodity. We should resist all criminal attempts to marketize, privatize, and corporatize water. Only through these means we can ensure the fundamental and inalienable right to water for the people all over the world.'

On 21 February 2004, the Government of Kerala declared Palakkad district to be drought affected, and ordered an immediate restriction on the company's usage of groundwater. On 9 March 2004, the company stopped operations. On 15 January 2005, marking the 1,000th day of picketing, protesters declared that the company would not be allowed to resume operations at any cost. At times, for example, between 8 and 15 August 2005, the plant operated, but eventually, it was shut down permanently.

The struggle in Plachimada against Coca-Cola not only raises issues of mindless destruction of groundwater by a multinational company but also exposes the gross inadequacies in the laws of governance and the rights to water. It also exposes the inability of political democracy, as we know it, to address weaknesses in law and governance. The state government meekly surrendered the opportunity that the struggle bestowed at the altar of the judiciary, rather than take advantage of it for bold political decision-making. The Plachimada struggle calls for the recovery of the Commons by communities.

GROUNDWATER RECHARGE CAN MAINLY BE DONE THROUGH

URBAN AREAS

All of us who directly consume water are the most important stakeholders in managing water. While many of us living in urban areas use or waste a lot of water, we rarely make an effort to conserve it. Fortunately, the rainwater harvesting method has provided a solution that can be practised easily in every household. It is a simple model where the roof acts as a catchment for rainfall, which, after flowing through a series of filters and pipes, is stored in ground-level containers for direct use or recharged into groundwater. An area of 1,000 square feet with one inch of rainfall is estimated to yield over 2,000 litres of water. Here are some popular methods for rooftop rainwater/storm runoff harvesting:

i. RECHARGE PIT

In alluvial areas where permeable rocks are exposed on the land surface or at very shallow depths, rooftop rainwater harvesting can be done through recharge pits. The technique is suitable for buildings having a roof area of 100 square metres, where pits are constructed for recharging the shallow aquifers.

ii. RECHARGE TRENCH

Recharge trenches are suitable for buildings having a roof area of 200–300 square metres and where permeable strata are available at shallow depths. The trench may be 0.5–1 metres wide, 1–1.5 metres deep, and 10–20 metres long, depending upon the availability of water to be recharged.

iii. TUBE WELL

In areas where the shallow aquifers have dried up, and existing tube wells are tapping deeper aquifers, rooftop rainwater harvesting through existing tube wells can be adopted to recharge the deeper aquifers. PVC pipes of 10 cm-diametre are connected to roof drains to collect rainwater.

iv. RECHARGE WELL

In areas where the surface soil is impervious and large quantities of roof water or surface runoff are available within a very short period of heavy rainfall, the use of trenches/pits is made to store the water in a filter media and subsequently recharge to groundwater through specially constructed recharge wells. This technique is ideally suited for areas where a permeable horizon is within three metres below ground level.

RURAL AREAS

In rural areas, rainwater harvesting is taken up considering the watershed as a unit. Surface spreading techniques are common since space for such systems is available in plenty and the quantity of recharged water is also high. The following techniques may be adopted to save water going to waste through slopes, rivers, rivulets, and nalas. The wisdom of water-scarce communities, such as the people of Rajasthan, is unparalleled: Ancient stepwell systems, nalas, and water gullies are a few such techniques. Here are some common practices in rural areas for water conservation and water harvesting:

i. GULLY PLUGS

These are built using local stones, clay, and bushes across small gullies and streams running down the hill slopes, carrying drainage to tiny catchments during the rainy season. Gully plugs help in the conservation of soil and moisture.

ii. CONTOUR BUND

These are suitable in low rainfall areas where monsoon runoff can be impounded by constructing bunds on the sloping ground all along the contour of equal elevation. Flowing water is intercepted before it attains the erosive velocity by keeping a suitable spacing between bunds.

iii. GABION STRUCTURE

This is a kind of check dam commonly constructed across small streams to conserve stream flows with practically no submergence beyond the stream course.

iv. PERCOLATION TANK

This is an artificially created surface water body, submerging in its reservoir a highly permeable land so that surface runoff is made to percolate and recharge the groundwater storage.

v. CHECK DAM/CEMENT PLUG/NALA BUND

These are constructed across small streams with a gentle slope. The site selected should have sufficient thickness of permeable bed or weathered formation to facilitate recharge of stored water within a short span of time.

vi. RECHARGE SHAFT

This is the most efficient and cost-effective technique to recharge an unconfined aquifer overlain by poorly permeable strata. The recharge shaft may be dug manually if the strata are of a non-caving nature. The diametre of the shaft is normally more than 2 metres.

vii. DUG WELL RECHARGE

Existing and abandoned dug wells may be utilized as recharge structures after cleaning and desilting them.

Water Conservation Starting from Home

1. BATHROOM AND TOILET HABITS

i. CHECK YOUR TOILET FOR LEAKS

Add a few drops of food colouring in your toilet tank. If, without flushing, the colouring begins to appear in the bowl, then you have a leak. A constant trickle into the toilet bowl wastes around 9,000 litres annually. To check for this type of leak, simply place a sheet of toilet paper under the rim in the back of the bowl (do this half an hour after the last flush when the bowl should be dry). If the paper soaks up water, you have a leaking cistern that may be wasting more than 375 litres of water a day.

ii. CHECK FAUCETS AND PIPES FOR LEAKS

Even a small drip can waste nearly 200 litres of water a day.

iii. PUT A PLASTIC BOTTLE IN YOUR TOILET TANK

Put an inch or two of sand or pebbles in the bottom of a one-litre plastic bottle to weigh it down. Fill the rest of the bottle with water and put it in your toilet tank, safely away from the operating mechanism. In an average home, the bottle may save nearly 20 litres of water every day without harming the efficiency of the toilet. If your tank is big enough, you may even be able to put in two bottles.

iv. STOP USING YOUR TOILET AS AN ASHTRAY OR WASTEBASKET

Every cigarette butt or tissue you flush away also flushes away over 20 litres.

v. TAKE SHORTER SHOWERS

A typical shower uses 20–35 litres of water a minute. Limit your showers to the time it takes to soap up, wash down, and rinse off.

vi. INSTALL WATER-SAVING SHOWERHEADS OR FLOW RESTRICTORS

Your hardware or plumbing supply store stocks inexpensive showerheads or flow restrictors that will cut your shower flow to about 11 litres a minute instead of 20–37 litres. They are easy to install, and your showers will still be cleansing and refreshing.

vii. TURN OFF THE WATER WHILE BRUSHING YOUR TEETH

Before brushing, wet your brush and fill a glass for rinsing your mouth.

viii. TURN OFF THE WATER WHILE SHAVING

Fill the bottom of the sink with a few inches of warm water in which to rinse your razor.

2. WASHING AND CLEANING

i. USE YOUR WASHING MACHINE AND DISHWASHERS ONLY WHEN FULLY LOADED

Remember that every time you run your dishwasher, you use over 90 litres of water, and your automatic washer uses 110 to 130 litres per cycle.

ii. IF YOU WASH DISHES BY HAND, DON'T LEAVE THE WATER RUNNING FOR RINSING

If you have two sinks, fill one with water to rinse. If you have only one sink, first gather all your washed dishes in a dish rack, then rinse them quickly with a spray device or a pan of water.

iii. DON'T LET THE FAUCET RUN WHILE YOU CLEAN VEGETABLES

Rinse your vegetables in a bowl or sink full of clean water.

iv. DON'T RUN THE HOSE WHILE WASHING YOUR CAR

Soap down your car from a bucket of soapy water. Use a hose only to rinse it off.

v. USE A BROOM TO CLEAN DRIVEWAYS, SIDEWALKS, AND STEPS

Using a hose wastes hundreds and hundreds of litres of water.

3. GARDENING AND LAWN MAINTENANCE

i. WATER YOUR LAWN ONLY WHEN IT NEEDS IT

Watering on a regular schedule doesn't allow for cool spells or rainfall which reduces the need for watering. Step on some grass. If it springs back up when you move your foot, it doesn't need water.

ii. WATER DURING THE COOL PARTS OF THE DAY

Early morning is better than dusk since it helps prevent the growth of fungus.

iii. PUT A LAYER OF MULCH AROUND TREES AND PLANTS

Mulch slows the evaporation of moisture.

iv. PLANT DROUGHT-RESISTANT TREES AND PLANTS

Many beautiful trees and plants thrive without irrigation, such as aloe vera, bougainvillea, khejdi, saangri, ashoka, and jacaranda, among others.

v. DEEP-SOAK YOUR LAWN

When you do water your lawn, water it long enough for it to seep down to the roots where it is needed. A light sprinkling that sits on the surface will simply evaporate and be wasted.

4. OTHER PARTS OF THE HOME

i. USE THE FAN-ONLY SETTING ON YOUR AC

Evaporative air conditioners can use around 100 litres of water per hour, adding up to 30,000 litres or more in a year. Switching your AC to the 'fan-only' setting for two hours a day, at night and on humid days, will make a big difference in your water use.

ii. REGULARLY CHECK FOR LEAKS, MAINTAIN FLOW CONTROL, AND CLEAN FILTER PADS

Comparing your evaporative air conditioning water consumption during periods of air conditioning use and non-use will help identify if there are any large water losses due to faults or leaks.

iii. USE EVERY DROP

Never pour water down the drain when there may be another use for it. Use it to water your indoor plants or garden. Learn to repurpose water. One easy way is to capture under your colander the potable water you use to rinse fruits and veggies, and deposit it in the garden. Do the same while you wait for your hot water to come in; don't waste the cold water as you wait for it to heat. Collect it in a bucket and use it to water your potted plants and garden. You could be catching as much as 10 litres for every minute you are waiting for the water to heat up.

iv. DON'T DEFROST WITH WATER

Instead of running water over food to thaw it, put it in the refrigerator to defrost overnight, or else use a microwave.

v. MONITOR YOUR WATER BILL FOR UNUSUALLY HIGH USE

Your bill and water metre are tools that can help you discover any leaks.

5. PARTICIPATE IN CAMPAIGNS TO STOP PRIVATIZATION OF WATER SUPPLIES

Almost 71 percent of Earth is covered with water. Seventy-five percent of the human body is water. And yet somehow we as a society find it difficult to get safe water to drink. Water remains one of our most scarce and sought-after natural resources. Some of the biggest conflicts between states and nations today are over it. Water is the next petroleum or 'blue gold' as they call it.

When we talk about privatization of water supply, we must know what it comprises. It is not just about supplying fresh potable water to people in urban dwellings. The urban water sector also involves multiple layers, including sourcing of water, deciding which is the best among the available options, getting potable water through purification plants for equitable distribution through huge infrastructure, along with managing the sewage generated through another set of huge infrastructure.

The private sector works on one bottom line: profit maximization. But the management of water supply is an issue of rights and a basic need, as acknowledged by the judiciary. Moreover, water is embedded in the ecosystem. Any attempt to see water only as a commodity is bound to have multiple disruptive consequences. Water is an asset of society and cannot be owned by the government, let alone the private sector. The government is the custodian of water on behalf of the people it represents. A failure in governance does not mean we sell the Commons to the private sector and make the fundamental right of people and other species a commodity to be profited from. It is important that all of us be vigilant and engage with the local municipality to hold them accountable and make sure the system works rather than taking the lazy way out and letting governments sell the supplies to the private sector.

WISDOM OF RAJASTHAN'S TRADITIONAL WATER HARVESTING SYSTEMS – EXPLORING THE WORKS OF ANUPAM MISHRA

Anupam Mishra was an Indian author, journalist, and Gandhian, as well as an environmental activist who worked extensively on water conservation. His study and work on the traditional water systems of Rajasthan make a strong argument for localism and cultural diversity, as well as for reclaiming and rejuvenating water as Commons as a means to ensuring resilience and sustainability as well as for social and cultural enhancement. He came into this field quite by accident. As a social activist, he arrived in **Bhinsar**, a small village in Bikaner district as an act of solidarity and support. Wandering in the pastures, Mishra saw a garden. Entering it, he saw a spotlessly clean courtyard surrounded by a wall. As he wanted to walk around, he was asked to take off his shoes (commonly done in temples and religious sites). In one of the corners, he saw a structure covered with a wooden lid, attached to it was a bucket tied with a rope. Opening the lid, he discovered water. It was a *kuin*, a water reservoir that collected rainfall; the courtyard was actually an *agor*, a catchment area of rainfall (Mishra, 2001: 105). This first encounter led him to study, protect, and promote traditional water systems. His first book on water traditions and practices (1993), published in Hindi as a copyright-free volume, was praised for empowering farmers, enabling them to be self-reliant. According to Mishra, 'the people of Rajasthan scaled the peaks of trade, culture, art, and standard of living because of the depth of their philosophy of life. This philosophy gave a special space to water. Although new development strategies have somewhat altered this exceptional water tradition, they have not been able to completely destroy it.'

The traditions of water management, which Mishra describes, continued for many hundreds of years. Until the nineteenth century, most resources were Commons, built or maintained and managed by local communities. They maintained their water infrastructure in three ways: collecting money, donating food grains, and labour. Everyone contributed as much as they could. In Jaisalmer, the task of cleaning and maintaining the lake was under the responsibility of the king himself and was done in a creative way. An invitation to play *lhas* (game) at Gharsisar Lake was sent to all the inhabitants of the city. When the day arrived, the king, his family, army, and subjects gathered at the lake. The king was the first to start digging, and soon everyone joined him. Music was played, and food was served to everyone who came to the lake. It was a playful event, conceived as a game, where the simplest peasant could rub shoulders with the mighty king – a case of 'help of all, help for all'.

To know more about this fascinating tradition of water management and harvesting, read Anupam Mishra's *Radiant Raindrops of Rajasthan*.

6. SAY NO TO CHEMICAL PESTICIDES

TENDING THE GARDEN OF HOPE

It has been nearly a month since you planted your Garden of Hope, and now you see a burst of activity and little tender plants sprouting from the earth. Your joy knows no bounds as you admire your precious plants which, through good care and proper maintenance, will blossom into your harvest of hope.

Watering the Garden of Hope is very important in the growing stages of the plants. Keeping a few basic concepts in mind can make it less time-consuming and more efficient.

i. Remember that high temperatures, low humidity, wind, and the sun contribute to the **evaporation** of water. Water droplets from a sprinkler head or hose nozzle may evaporate before they touch the soil. To minimize evaporation, water the garden when the air is still, in the early morning hours or during the cool evening (don't water the leaves to prevent the spread of disease). Use a nozzle that produces large droplets rather than a fine spray. As plants grow, they need more water because more leaves equals more loss of water from the leaves. This is known as **transpiration.**

ii. As the season progresses, most plants will need more water. However, if too much water is given, soluble fertilizer can be leached from the soil. When plants are small, only the soil around their root systems should be kept moist; towards the end of the growing season, soil down to one foot or more should be kept moist. Avoid frequent, shallow watering because this hinders deep root development. Sprinkle water uniformly and slowly to prevent runoff and erosion. If in doubt about when to water, check your soil by sampling it: Dig down carefully, making a narrow, deep hole one foot or more away from the plant and about 4–12 inches deep to where the plants' roots grow. Squeeze some soil into a ball in the palm of your hand. If it holds together, there is enough moisture. If it crumbles, it is time to water. If the hose has been lying in the sun, let the water run for a few seconds on a stone or on a mulched area until the water comes out cool.

The layout of the patch and the positioning of the pots also make a world of difference. Plenty of breathing space and a free flow of nutrients are necessary for the plants. Weeds are plants that may compete for nutrients from the soil. They can restrict the necessary nutrients from reaching the plants. Some weeds can also grow larger than the plants and not let in sunlight. Therefore, to ensure the health of plants, we need to remove weeds when we see them. Regular checks, surveys, and monitoring are good ways to get to know the nature of your plants and their surroundings. Talking to and touching the plants has proven to be healthy for their well-being.

Tending and caring for your plants is vital. The adage 'You reap what you sow' implicitly suggests the nurturing and patience required between the sowing and reaping. But there are certain challenges that you may have to face before the harvest reaches its fruition. They will come in the form of pests and weeds. These will be the defining moments as you struggle in your mind whether to use chemicals to save your plants, or stick by the principles of organic farming and let Mother Nature look after your plants naturally. As your guide and friends from Navdanya, we have put together some simple natural remedies from nature's kitchen to help you fight pests and weeds in your garden.

Insects and pests also require our attention, as some can be harmful to the plants. Usually, insects like butterflies, beetles, caterpillars, and spiders positively impact your garden's health. However, occasionally, a harmful insect or pest can also take an interest in your patch. Flowers are the best way to distract insects' attention. Instead of using chemicals and poisons to quickly eliminate the problem, study and analyse it. There are many natural ways to control harmful pests and diseases. Weak solutions of neem, garlic, vinegar, mint, or chilli oil sprayed onto the leaves deter aphids and whiteflies. Feeding beneficial bacteria (molasses, cornmeal) into the soil can also make the plants healthier and more resilient. There are also some who believe milk to be the cure-all solution. A weak solution of milk or baking soda in water, when sprayed on leaves, is believed to have a great effect on mould and fungal problems.

CITIZEN ACTION IN DELHI AGAINST WATER PRIVATIZATION – A VICTORY FOR THE PEOPLE

On 13 June 2005, five farmers were shot dead in Tonk, Rajasthan, during a protest. They were demanding their share in the water from Bisalpur Dam, which was diverting water from villages to the city of Jaipur under an ADB project for water sector 'reforms' in the state, then ruled by the BJP government. Sonia Gandhi, the then president of the Congress Party, rushed to Tonk, called the firing barbaric and offered relief to the affected families. Meanwhile, the Congress government in Delhi was determined to create another Tonk in Muradnagar, Uttar Pradesh, with its demand to divert 635 million litres of Ganga River's water per day to the Sonia Vihar Water Treatment Plant (WTP). The WTP had been privatized to the French company, ONDEO-Degrémont, a subsidiary of SUEZ-Lyonnaise des Eaux water division, the water giant of the world.

The government was acting on the World Bank's recommendation and other aid agencies having a history of creating water markets for global water MNCs while robbing the Indian people, both hydrologically and financially.

Delhi has been sustained for centuries by the Yamuna River. Decades of industrialization have turned the Yamuna into a sewer and toxic drain. Instead of stopping the pollution, using the scarcity created by the pollution, the World Bank started to push the Delhi government to privatize its water supply and get water from the Tehri Dam on the Ganga, hundreds of miles away.

The privatization of Delhi's water supply is central to the Sonia Vihar WTP. Inaugurated on 21 June 2002 by the chief minister of Delhi, the plant was designed for a capacity of 635 million litres a day on a 10-year BOT (build-operate-transfer) basis, at a cost of 1.8 billion rupees (approx. 50 million dollars). The contract between Delhi Jal Board (the Water Supply Department of the

Delhi Government) and ONDEO-Degrémont was supposed to provide safe drinking water to the people of New Delhi.

Navdanya, along with the Right to Water Campaign, Citizens' Front for Water Democracy, Delhi Water Sewer & Sewage Disposal Employees' Union at the Delhi Jal Board, the Water Workers' Alliance, and countless others, conducted relentless public education and protest campaigns against this violation of the fundamental right of access to water for all. Through this campaign, residents of Delhi were educated on the impact of water privatization, and they were asked to put pressure on the Delhi Government, Central Government, and the World Bank country representatives.

Cross-class alliances were forged between the middle-class residents of posh residential colonies and poor residents of slum clusters. A common set of concerns and demands was constructed after several rounds of meetings to ensure that everyone could come together on a common platform to oppose privatization plans under the guise of providing water 24 hours a day, seven days a week. The residents of Delhi were determined not to allow one group to be played against another. Finally, on 23 November 2005, the Delhi government wrote to the Central Government saying it wished to withdraw its loan application to the World Bank. This was the culmination of a year-long battle by the people of Delhi, who organized under various organizations to fight the Delhi government's rapidly evolving programme to privatize the city's water supply system without any public consultations. The residents of Delhi did not rest easy even after this historic win. They got together with water and management experts from across the country and sent a list of suggestions to the Delhi government on 5 December 2005. The suggestions were to install water metres and to make the system transparent so that people could see online how much water was flowing into each one of the city's 21 zones. This would enable those who got step-motherly treatment to question their public representatives.

Natural Solutions to Control Pests

If intercropping is not sufficient to keep away the pests, you can also spread a natural mixture of products on the field that drives the annoying insects away.

Neem is commonly used as a natural pest control: the seeds as well as its leaves. Tobacco leaves, custard apple leaves, walnut leaves, garlic leaves, coriander leaves, onion leaves, and vitex leaves are used to eradicate pests in case neem is not available. Soak the leaves in water overnight. The next morning, take out the leaves and pound them. Put them back in the water and leave for an hour. Filter the water and use it as a pesticide. The residue can be combined with compost or cow dung. You can spread this pesticide on your garden. However, do not repeat for two or three weeks.

Chop or grind an onion and a pod of garlic. Add 1 teaspoon red chilli powder with the onion and garlic to a litre of water. After an hour, strain the liquid and add 1 tablespoon of any dish soap. Spray your plants with this liquid, covering both sides of the leaves. You can refrigerate and store this mixture for up to a week. This is an excellent way to stop leaf-eating insects in their tracks!

Weeding is a job few people like unless it is a team effort. Mulching helps, but some weeds will persevere and need to be removed by hand. The key here is to be just as **persistent** in scouting and pulling weeds as they are persistent in finding spaces to invade. The proud and avid gardener goes out into the garden frequently to check the plants' progress, scout for weeds and insects, see if the plants need water, and enjoy being outdoors.

Some vegetables need to be trained to upright structures for the best results. These include tomatoes that take up less room when supported by cages. Cucumbers, pole beans, and peas can also be trained on fences, trellises, or poles. Get this task out of the way early since it is difficult to untangle plants once they are older.

Keen observation can save your garden from pests and weeds as well as increase your concentration and help you enjoy nature more, so keep observing and helping your garden grow!

READING LIST

ANUPAM MISHRA, *The Radiant Raindrops of Rajasthan*, translated by Maya Goburdhun. The Research Foundation for Science, Technology and Ecology (2009).

VANDANA SHIVA, *Water Wars: Privatization, Pollution, and Profit*. Penguin Publication (2007).

CHARLES FISHMAN, *The Big Thirst: The Secret Life and Turbulent Future of Water*. Simon & Schuster (2011).

Recipes to Protect Water as a Commons

AMARANTH CUTLETS

INGREDIENTS

- 200 gm potato
- 15 gm carrot
- 15 gm peas
- Fresh coriander leaves
- 1 green chilli
- 75 gm popped amaranth/ramdana/rajgira
- 3 tsp garam masala
- 2 tsp black pepper
- 50 gm peanuts
- Salt, to taste
- Oil, for frying

METHOD

- Boil, peel, and mash the potatoes and carrots. Boil the peas separately. In a medium-sized bowl combine the boiled peas with the potato and carrot mash.
- Finely chop the fresh coriander and green chilli. Keep aside.
- Add the popped amaranth, chopped coriander, green chilli and all the other spices to the bowl. Season to taste with salt. Mix well to combine all the ingredients. The dough is now ready.
- Oil your hands and taking a portion of the dough, make oval-shaped cutlets; press them between your palms to flatten.
- Heat the oil in a deep frying pan over medium heat. In batches, put the cutlets in the pan and fry until they turn golden brown.
- Use a slotted spoon and drain the cutlets on kitchen towels.
- Serve hot with a chutney of your choice.

A hardy crop, Amaranth requires very little water and gives a very high and nutritious yield of both fibre and grain. It is an important crop in China and Russia, too.

Source: Bhoole Bisre Anaj/Forgotten Foods, *Navdanya*

COOLING CURD RICE *for the* STOMACH

INGREDIENTS

2½ cups curd/yogurt
4 green chillies, chopped
1 small piece ginger, minced
1 tsp salt
½ cup cooked rice
2 tsp ghee
½ tsp mustard seeds
1 tsp urad dal/split black gram
1 tsp chana dal/Bengal gram
1 sprig curry leaves
Fresh coriander leaves for garnishing

METHOD

- In a large bowl, combine the curd with green chillies, minced ginger, and salt. Whisk well.
- Now add the cooked rice to this curd mixture.
- Prepare the seasoning by heating the ghee in a small pan over low heat. Add the mustard seeds, urad dal, chana dal, and curry leaves. Cook until the lentils splutter. Add the seasoning to the mixed curd and rice. Combine well.
- Garnish with fresh coriander and serve.

Source: Akshat, *Navdanya*

GO COLA-FREE *with* YUMMY INDIGENOUS **THANDAI**

INGREDIENTS	METHOD
½ cup almonds ½ cup fennel seeds 50 gm melon seeds 50 gm poppy seeds 20–25 black peppercorns 30–35 green cardamom 2½ cups water 4 cups sugar 3 tbsp rose water	• Soak the almonds, fennel seeds, melon seeds, poppy seeds, peppercorns, and cardamom in water, in separate bowls for 2 hours. • In a tall flask, combine the water and sugar and keep aside. • Remove any excess water from all the ingredients with the help of a sieve. • In a mixer-grinder, grind all the ingredients with the sugar syrup to a fine paste. • Add the rose water. • In a glass of milk, add 2½ tbsp of the prepared thandai mix and top it with few ice cubes. • You can store the thandai mix in a clean glass bottle.

(Recipe courtesy: Nisha Madhulika, retrieved from Kavita Mahindra's post in Sikandalous Cuisine, featured in Biodiversity in a Glass *by Navdanya)*

5

SLOW HEALTH

Nutrition, the Planet, *and* You

'We are caught in an inescapable network of mutuality, tied in a single garment of destiny. Whatever affects one directly, affects all indirectly.'
DR MARTIN LUTHER KING JR

Our planet's health and our health are not separate from each other. The illusion that plants and animals are machines for manufacturing raw materials, which become fuels for our bodies, which are also machines, has created the industrial agriculture and food paradigm. This approach to our bodies is at the root of the current explosion of chronic diseases.

As forests are destroyed and our farms become industrial monocultures to produce toxic, nutritionally empty commodities, and our diets become degraded through industrial processing with synthetic chemicals and genetic engineering, we get connected to each other through disease. Instead of being connected through biodiversity within us and outside us, through a continuum of health, we can be linked worldwide through the spread of diseases like the Covid pandemic in 2020, when we invade the homes of other species, manipulate plants and animals for commercial profits and greed, and spread monoculture. Or we can be connected through health and well-being for all by protecting the diversity of ecosystems and protecting the biodiversity, integrity, self-organization (autopoieisis) of all, including humans.

Over the past 50 years, 300 new pathogens have emerged as we destroy habitats and manipulate them for profit. Prominent examples include Ebola, which has been linked to rapid deforestation;[1] Kyasanur Forest Disease (KFD), a highly pathogenic virus that spread from monkeys to humans through virus-infected ticks as deforestation shrunk the forest habitat of monkeys[2] in South Canara, India; and 'mad cow disease' or bovine spongiform encephalopathy[3] (BSE), a disease that affects the brains of cattle and infects humans with Creutzfeldt-Jakob disease (CJD), a degenerative brain disorder that leads to dementia. When animals are manipulated, and their integrity and right to health is violated, new diseases can and do emerge. Earth's biodiversity richness in the forests, in the farms, in the food we eat, and in our gut microbiome makes the planet, her diverse species, including humans, healthier and more resilient to pests and diseases.

The health emergency that the Covid pandemic woke us up to in 2020 is connected to the crisis of extinction and disappearance of species as well as the climate emergency.

All emergencies are rooted in a mechanistic, militaristic, anthropocentric worldview of humans as separate from, and superior to, other beings who we can own, manipulate, and control. It is also rooted in an economic model based on the illusion of limitless growth and greed, which systematically violates planetary boundaries and ecosystem and species integrity.

With the most recent pandemic and global health crisis, we need to look at systems that spread disease and structures that create health in a holistic, systems approach. A systems approach to healthcare in times of a pandemic would address not just the virus, but also how new epidemics, comorbid conditions related to non-communicable chronic diseases are spreading due to an unsustainable, anti-nature, unhealthy industrial food system.

The health consequences of maladapted modernity, driven by commercial food systems, are currently being experienced in epidemic proportions across the world. Apart from premature death and prolonged disability, diseases resulting from poor diets are forcing people to seek expensive healthcare. The global costs of healthcare due to food system-related illness are soaring, with obesity set to cost $1.2 trillion by 2025, diabetes $1.5 trillion by 2030, and cancer $2.5 trillion.[4]

The indigenous systems of healthcare have been criminalized by colonization and the pharmaceutical industry. The basis for most of them was the food we ate. As Hippocrates said, 'Let food be thy medicine and medicine be thy food.' Thus, what we eat and how much of it affects our overall health and mental well-being. It is estimated that 80 percent of disease can be tied to our food and lifestyle choices.

Today, most of us use pills to sort out every aspect of our health, from vitamin pills, to fiber pills, to anti-inflammatory pills; the list goes on and on. However, if we pay attention to our life, our body and mind, and the wisdom of our ancestors, what in India we call 'nani-dadi ke nuskey', in other words, the antidotes given by our grandmothers, we may be well on our way to good health and also create an abundance of health for our planet.

Slow health is an understanding that our planet and our bodies have the ability to mend themselves if we look after **Mother Earth**, we look after one another, and most of all, after ourselves.

---

1. https://newint.org/features/web-exclusive/2018/04/10/deforestation-ebola-outbreak
2. https://www.deccanherald.com/state/deforestation-behind-kfd-713955.html
3. https://www.emedicinehealth.com/mad_cow_disease_and_variant_creutzfeldt-jakob/article_em.htm
4. https://www.business-standard.com/article/current-affairs/global-cost-of-treating-obesity-related-illnesses-to-top-1-2-tn-from-2025-117101000472_1.html

Slow Health vs. Fast Health

'We are now pretty clearly involved in a crisis of health, one of the wonders of which is its immense profitability both to those who cause it and to those who propose to cure it. That the illness may prove incurable, except by catastrophe, is suggested by our economic dependence on it.'
WENDELL BERRY
Health is Membership, 1994

The Hopi (Native American tribe) describe the phenomenon of destroying everything that sustains a society as 'Powaqqatsi' or parasitical life. In this hyper-consumerist highway that we are speeding on, corporations keep extracting from and degrading the planet and impoverishing its soils and citizens; our fragile web of life is poisoned and broken.

The word 'health' comes from the same Indo-European root as 'heal', 'whole', and 'holy'. To be healthy is literally to be whole; to heal is to make whole. However, today 'healthcare' is about specialization and is in fragmentation, with technology and drugs at its centre, and the human body, just like a machine, where parts are tinkered with and adjusted periodically to keep it running. When we distance ourselves from individual parts of our own body and isolate the treatment to different parts, a separation from the 'whole' or health takes place. How can this separation then truly heal us or make us whole again? The art of slow health recognizes that rest, food, and ecological health are the way to 'healing', to becoming whole. Are we confronting some fundamental incompatibility between mechanical efficiency and organic health? What are the fundamental differences between fast treatment and slow health?

FAST TREATMENT/ HEALTH	SLOW HEALTH
Specialized Looking at the particular body part and treating it with little understanding of how we got here and what all needs to change to ensure one's good health.	**Holistic** Looking at the whole body, the community we live in, and our natural environment to build resilient and healthy communities and people.
Separation Seeing the mind, body, and spirit as separate from each other, and humans as separate from other species or ecology.	**Integration** Looking at the interconnected relationship between mind, body, and spirit, and linking it to the health of our planet and others.
Food as Poison There is a silent pandemic of food-related lifestyle diseases such as obesity, diabetes, hypertension, and cancer spreading through a systematic, methodical process of marketing, addictive additives, and government and transnational nexuses operating as a poison cartel, pushing us towards pharmaceutical solutions.	**Food as Medicine** Eating fresh, non-processed, local and ecological food, free of pesticides builds immunity and holistic health, from the mind to the body. Moreover, if you start your vegetable patch, then the connection to soil fuels the soul as well.

Spasmodic A greater emphasis on curative medication, looking at the body as a machine, wherein one needs to go to a mechanic when things need to be fixed.	**Continuum** Everyday mindfulness and observation of the body, the quality of ecological wholeness for the community in terms of clean water, air, and soil, as well as curative inputs for when we need a little help to feel better.
Knowledge Colonization Criminalization of indigenous knowledge and discrediting ancient wisdom as unscientific.	**Knowledge Sovereignty** Recording the wisdom of our ancestors and learning from them about the techniques of mind, body, and spirit connection.
Knowledge Monopoly Patenting life-saving medication and exploiting indigenous wisdom on health for profit.	**Knowledge Diversity** The health cultures of communities range from food that heals, to fasting habits and seasonal eating, creating community celebration and health for all.

Poison or Food

'When diet is wrong, medicine is of no use. When diet is correct, medicine is of no need.'
AYURVEDA PROVERB

For the last few decades, there has been an exponential rise in non-communicable chronic diseases that are killing people. Toxics and industrial food systems are major contributors. We must remember that food is not a commodity; it is not 'stuff' put together mechanically and artificially in labs and factories. Food is life.

However, not only are our plants getting less nutritious, but they are also getting more toxic. As food writer Mark Bittman remarks, since food is defined as 'a substance that provides nutrition and promotes growth' and poison is 'a substance that promotes illness', then 'much of what is produced by industrial agriculture is, quite literally, not food but poison'. As the unabashed use of biocides (herbicides, insecticides, fungicides, as well as synthetic chemical fertilizers and antibiotics) kills or disrupts soil microorganisms that allow plants to absorb nutrients, the increase of atmospheric CO_2 is leading to accelerated photosynthesis. Plants grow faster but contain fewer nutrients.

Industrial activities – including mining, smelting, and manufacturing; domestic livestock and municipal wastes; pesticides, herbicides, and fertilizers used in agriculture; petroleum-derived products that are released into or break down in the environment; and fumes generated by transportation – all contribute to poisoning of soil, thus poisoning our food. Another growing concern is the 'emerging pollutants' that include pharmaceuticals, endocrine disruptors, hormones, biological pollutants, 'e-waste' from old electronics, and plastics, which are part of almost every human endeavour.

The court cases in recent years have showcased the links between Roundup pesticide and cancer. With the buildup of liabilities related to cancer cases, thousands of such cases linked to glyphosate have been filed in the United States. In the cases of Johnson Edwin Hardeman, Alva and Alberta Pilliod, the courts have ruled in favour of the cancer victims. Almost 10 million people die from cancer annually.

Every sixth death in the world is caused by cancer. Glyphosate is the most heavily applied herbicide in history, globally. It has been found in air, rainwater, drinking water, and in groundwater used for irrigation. It moves through leaves, stems, seeds, roots, and soil. It moves through human organs, lymph tissue, and blood. It poisons microbiome. One of the best-known glyphosate-based products is Monsanto's Roundup weed killer. A US federal court in 2018 ordered biotech giant Monsanto to pay a compensation of $289 million to Dewayne Johnson, a ground-keeper at a school in San Francisco, for having got cancer after using glyphosate present in pesticides such as Roundup. To heal ourselves and the planet, we must hold the poison cartel responsible and act against them.

THE BIRTH OF POISON CARTEL FROM WAR INDUSTRY TO FOOD INDUSTRY

Monsanto is the world's largest producer of genetically modified (GMO) crops and maker of the glyphosate-based herbicide Roundup. Bayer, the transnational renowned for the creation of Aspirin (medicine for reducing fever and pain), has also been subject to criticism over its widely used insecticide, imidacloprid, which belongs to a controversial class of chemicals called neonicotinoids that's linked to widespread deaths of pollinators.

Monsanto, now merged with Bayer, has a long history of creating explosives and lethally poisonous gases using shared technologies and selling them to both sides in the two world wars. The same war chemicals were bought from the same manufacturers with money borrowed from the same bank.

MoBay supplied ingredients for Agent Orange in the Vietnam War. Around 20 million gallons of MoBay defoliants and herbicides were sprayed over South Vietnam. Children are still being born with birth defects; adults have chronic illnesses and cancers due to their exposure to MoBay's chemicals. *Monsanto and Bayer's cross-licensed Agent Orange resistance has also been cross-developed for decades.* Wars were fought, lives lost, nations carved into holy lands – with artificial boundaries that suit colonization and resource grab. The Monsanto-Bayer combination is an example of the rapidly consolidating agricultural industry, with only a handful of companies controlling the sector. Alongside slumping crop prices, DuPont Co. and Dow Chemical Co. have agreed to merge, as did China National Chemical Corp. and Syngenta AG.

As we stand today, the relationship between Big Pharma, Big Food, and Big Tech is *coalescing* by the minute and will seal the future of our health in a web of undemocratic and dangerous consumption patterns if we do not remain vigilant.

Eat 'Real', Eat Healthy: Create a Poison-free Future

'But man is a part of nature, and his war against nature is inevitably a war against himself.'
RACHEL CARSON

How we produce, distribute, and consume our food is increasingly becoming the centre of the multiple crises we face today. Whether it is ecological crises of biodiversity decline and climate change or the health crisis of hunger, malnutrition, and the spate of diseases, the root of the problem is a growing dependence on a dysfunctional paradigm that depends on pesticides and economies of scale to accelerate the quantities of food produced – all at the expense of nutritional quality. This causes a wide variety of detrimental impacts on health as well as the ecosystem. Evidence of the toxic consequences of the policies of agrichemical mega corporations, which we call the 'poison cartel', is growing by the day: Species are disappearing, small farmers are losing their livelihoods, rural populations are being driven from their land to make way for industrial agriculture, and consumers increasingly have no option other than to buy food grown in increasingly contaminated soils. There is a need to take steps both individually and through our local communities to build a poison-free future.

Here are some steps that can start you off on this journey:

1. GUT INSTINCT

Just as the soil requires diverse micronutrients and organisms to remain healthy, so does our stomach. The fact is that both these are interrelated, for a healthy stomach will indicate a healthy planet and vice versa. Just as your choices at the grocery store dictate your health, they also dictate the health of the Earth. Non-organic foods are sprayed with extremely toxic chemicals that kill everything other than the crop itself. This includes the living organisms that a plant needs to grow and make the soil nutrient-rich. So, notice your stomach health, and you will learn a lot about your health and our planet's health as well.

2. KNOW HOW TO SHOP FOR CHEMICAL-FREE PRODUCE

In the US, the Environmental Working Group (EWG) publishes two lists: the Dirty Dozen and the Clean 15. The EWG rankings are based on USDA-tested levels of chemical residues that remain on conventionally raised fruits and vegetables after washing. While no system of ready consumer information exists in India, we have put together a list of our dirty dozen. Here are some ways to spot what is toxic in your food produce:

i. IF IT HAS GOT THE LOOKS, LOOK AGAIN

To achieve an instant shine, vegetables such as cabbage **are dunked** in a carbofuran (furadon) dip to give them a greenish-blue sheen. The scientific literature handed out by the companies who produce it warns that any produce on which carbofuran is used should not be consumed for a minimum of 28 days. However, this is not normally the case. So, when buying cabbage, choose one that is dark green and without a sheen; the same applies to other fruits and vegetables. They may even look muddy. Just throw away the top leaves. Another staple in our kitchens is cauliflower. The harvested flower of this vegetable attracts the semi-looper worm and a dozen other caterpillars. We hate to see caterpillars in our cauliflower, so traders treat it with endosulphan. Vegetables treated with endosulphan should not be eaten for 14 days. To prevent the tiny black spots that affect cauliflower, farmers use fungicides like Bavistene. The small, fine black dots are actually not harmful to health. But since consumers want a resplendent white cauliflower (no matter what it looks like after it has been cooked!), the farmer is ever ready to oblige.

ii. NATURALLY SHINY SKIN PRODUCE DOUBLE TROUBLE

Vegetables like tomato and brinjal have naturally smooth and shiny skins, thus, external application of pesticide may not be effective. Farmers therefore use what are called **'systemic or man-made or factory-made insecticides' such as Roundup, urea etc.** In fact, most companies are now largely promoting systemic pesticides. These enter the plant through the stomata, contaminating the tomato from within. No amount of washing of the vegetable will get rid of it. According to Miguel Braganza, a former agricultural scientist with the Government of Goa's agriculture department and now additional director, Organic Farming Association of India (OFAI), these nerve poisons used on such vegetables were first used during the Second World War, where canisters of these nerve poisons were dropped into the tanks, disorienting the commanders, and forcing them to emerge in the open. When buying produce, look for a vendor who stocks different varieties of the same vegetable. This would indicate a diverse supply coming from small farms.

iii. RIPEN TO HURT

Fruits like mangoes are mass-produced and are transported great distances. They are often plucked before they are seasoned and then quickly ripened using ethylene gas. Ethylene gas is not harmful in itself. In fact, it is released naturally by the mango from the seed outwards as it ripens. But subjecting the fruit to this gas externally ripens the mango from the skin inwards. As a result, though the skin looks a golden yellow, the fruit is most likely to be sour. Another common yet dangerous fruit-ripening agent is calcium carbide because it is usually contaminated with arsenic, a deadly, poisonous heavy metal. So, look for organic varieties or know the farm you are sourcing them from.

iv. POULTRY PROBLEMS

'Battery farming' and a constant diet of artificial hormones are used to quickly fatten the poultry; that is, what would take nine weeks to achieve now takes just about five weeks! The growth hormones used are estrogen and progesterone. Young girls consuming chicken raised in this manner may experience puberty earlier than usual, or face possible damage to their follicles, or develop polycystic ovaries (PCOS). Poultry farmers also add a variety of antibiotics and vaccines in the food or water the birds consume. If you cannot find naturally raised country chicken, eat less meat.

NOW FOR SOME SAFER OPTIONS FOR YOUR GROCERY BAG

ROOT CROPS

The root crops, such as radish, carrot, and beetroot, but not potatoes, are generally safe, as they are not grown on as large a scale as potatoes. These are also hardy crops that don't require regular doses of pesticides and chemicals to increase their shelf life.

YAMS

These are also a safe source of good and nutritious food. Chemicals are not usually used for such crops. The same cannot be said of tapioca, which is now regularly sprayed because of insect attacks.

COCONUTS

The flesh is filled with vitamins and minerals. Its pH is alkaline, and can therefore be used in place of milk in case you want to offset a very acidic stomach. Some pesticides are used to get rid of beetles that tend to the tree, but normally not on the fruit.

GOURDS

The entire range of vegetables from the family of cucurbits is not generally sprayed, except for cucumber. Most of the hard-skinned ones, including bottle gourd, ridge gourd, snake gourd, and pumpkin are safe to eat because they are produced in kitchen gardens where dangerous chemicals are not normally used.

PAPAYA

The problems papaya faces cannot be really solved by the existing brand of chemicals. If you get local varieties of papaya (with seed), these are absolutely safe, since they have been grown in homesteads or home gardens. The more misshapen or unique the shape, the better the guarantee that you are having one that is not grown in large plantations and is more natural.

SAPOTAS

The chickoo plant is attacked by bats, for which chemical sprays are of no use. The other fruits that remain generally safe include pomegranates, oranges, and sweet limes, where very little chemical is used.

JACKFRUIT

Normally no chemicals are used at all for growing them. Your best choice, however, is to find food grown on an organic farm where regulations do not permit the use of either chemical fertilizers, or poisonous pesticides, or genetically engineered seed.

3. NURTURE POISON-FREE ZONES, COMMUNITIES, FARMS, AND FOOD SYSTEMS

A recent study by the University of California, Berkeley, and Friends of the Earth on what the 'pesticides in our urine tell us about organic food' found that 'after only six days on an organic diet, every single person would see significant drops in those pesticides [in their urine], including several linked to increased risk of autism, cancer, Parkinson's, infertility, and other significant impacts on health.' So, eating poison-free food becomes a health imperative today. Here are the small steps you can take to start this journey:

i. Reduce the amount of processed food in your diet.

ii. Family weekend & community cooking: If you are in a hurry and need quick fixes, set two days in a month aside to make your own sauces, curries, dips, jams, and chutneys. Make it a fun family 'weekend bender', or call friends and cook together, or **barter** food amongst your friends and family so that you can buy in bulk, reducing cost and ensuring safer food choices.

iii. Know your food: Learn about healthier food choices; create a WhatsApp group and share information with one another about healthy food options in your society, locality, or area.

iv. Plan your meals in advance: This is a small step that can help you buy produce smartly and also help you track the kind of diet you are currently consuming and how it impacts not only your health but the ecosystem as well.

4. CREATE AND SUPPORT LOCAL FOOD ECONOMIES, FARMERS' MARKETS, CSAS, BIO-DISTRICTS, ECO-ZONES

i. Source your produce from local farms that grow organic food. Buy in bulk to make it budget-friendly and don't forget to share with friends and family.

ii. If you have a favourite restaurant, ask them where they source their raw produce from and encourage them to use organic local produce.

iii. Encourage your office and school cafeterias to source organic produce and offer healthy, seasonal options. You can help by giving them contacts to source the produce from, and also suggest easy recipes for the menu.

iv. Build links between schools, hospitals, healthcare centres and local organic fresh, diverse food systems by organizing nutrition literacy workshops. The current trend of counting calories does not give you the full picture of what you are eating. You can get 100 calories from fruits or junk food; the impact on your health will be different, thus it is important that school and residential associations get involved and provide nutrition literacy to all.

5. GROW GARDENS OF HOPE, GARDENS OF HEALTH

According to research, people who spend time in a garden or are surrounded by plants are significantly more likely to experience good health, higher psychological well-being, and greater physical activity levels. Published in the journal *Landscape and Urban Planning*, the study analyzed data from nearly 8,000 people collected by Natural England between 2009 and 2016. It found the benefits of gardening to health and well-being were similar to the difference in health between people living in the wealthiest parts of the country compared to the poorest. The benefits applied to whether people spent their time gardening or simply relaxing. We have started your journey of growing a garden from the beginning of the book, and hopefully, you are reaching the point of harvesting your garden. At the end of this chapter, we will be discussing how to harvest your Garden of Hope. If you have not started yet, we hope you feel inspired now to begin your journey, however small or big. You can start with simple medicinal herbs like holy basil or aloe vera, which are easy to look after, and even grow in pots. Fast-growing greens like spinach, mint, and curry leaves are also now regularly sprayed with chemicals. (Sometimes farmers mix different pesticide formulations together to get a stronger 'cocktail'). Spinach is sprayed once a week to ensure that insects do not chew up some of the leaves, which induces the customer to reject the whole bundle. Monoculture plantations of curry leaves are also sprayed. The important thing to note about greens is that they are easy to grow in your own home, in your balcony, on the terrace, or in the garden.

6. CELEBRATE SEED FREEDOM, FOOD FREEDOM, AND EARTH FREEDOM

In India, we are blessed with many harvest festivals; however, mere rituals without a sound understanding of the philosophy and ethos of these festivals are of no use. Understand the importance and reasons behind these festivals, including the fasting rituals and the impact they have on your health. Mostly, the food eaten on these dates is an indication of the seasonal shift in diet and also lets you know what to consume in those seasons to enable health. Do a deep dive and learn more about these festivals and their links with your health. Eid ul-Fitr, Pongal, Navratri, and many more festivals have specific food associated with them; many detox and reboot our system to prepare for the seasonal changes, both in weather and vegetation.

Lifestyle Diseases, Creating Addictions, and Blaming the Victim

'It is important to the health of the nation that we remove the stigma associated with obesity. It is not a lifestyle choice caused by individual greed but a disease caused by health inequalities, genetic influences and social factors. It is governments, not individuals, which can have an impact on the food environment through regulation and taxation, and by controlling availability and affordability.'

PROFESSOR ANDREW GODDARD, RCP PRESIDENT

'Obesity should be recognized as a disease'

presented to RCP Council, 17 July 2018

Humans have an inherited preference for energy-rich foods such as fats and sugars. Food scientists know this and create ingredients that are far higher in fat and sugar than what is available through nature. The most common of such sugars is the intrinsically addictive high-fructose corn syrup (HFCS). In fact, foods that traditionally were not sweet, for instance pasta sauce, are now artificially sweetened to keep the craving high – often with sugar levels that can rival those found in packaged desserts. HFCS is found in everything, from ketchup and salad dressing to cereal and breads.

As an ingredient, HFCS was shown in many studies to be as addictive as drugs, like cocaine or heroin, with salt proven to have similarly addictive, opioid-like qualities. Having that sugar could release dopamine, the feel-good hormone in the brain, which associates that food with pleasure, causing the body to crave more. This feel-good sensation keeps one hooked on certain foods, bringing instant comfort and satisfaction.

A popular potato chip advertisement has the catchline 'No one can eat just one'. And they are right. Many common and popular snacks have been expertly engineered to keep us addicted, almost constantly craving more of whatever falsely satisfying manufactured treat is in front of us. One large-scale study, involving more than 100,000 adults, found that eating 10 percent more ultra-processed foods was associated with an increase of above 10 percent in the risks of cardiovascular disease, coronary heart disease, and cerebrovascular disorders.[1] A sugar-laden diet may raise your risk of dying of heart disease even if you aren't overweight.

Salt, too, is extremely addictive, just as much as sugar. It doesn't matter if the salt is white, pink, sea salt, or crystallized – all of it has the same effect on one's body. Though packaging may lead you to think that certain salts are healthier, they are all the same – salt, which is bad in excess.

Even if a food isn't overly salty, salt may sneak into packaged food more rampantly than expected. In most cases, salt is used as a preservative to give food an extended shelf life and keep it safe. Salt can also be used to enhance a food's colour (such as making the crust of bread a more appealing golden brown), as well as a flavour enhancer in foods you may not associate with saltiness, like ketchup or breads.

Beyond overeating in general, eating too much salt is proven to have negative effects on human health. The more salt you eat, the higher your blood pressure; this can put a strain on your heart, kidneys, brain, and arteries, often leading to a stroke, heart attack, or kidney disease.

In the last few decades, ultra-processed food intake has increased dramatically worldwide. These foods now account for 25–60 percent of a person's daily energy intake throughout much of the world. Chemically processed foods often only contain refined ingredients and artificial substances with little nutritional value. They tend to have added chemical flavouring agents, colours, and sweeteners.

Some examples of ultra-processed foods include: frozen or ready-to-eat meals; baked goods, including pizza, cakes, and pastries; packaged breads, cookies, biscuits; processed cheese products; breakfast cereals, crackers and chips, candy and ice cream; instant noodles and soups; reconstituted meats, such as sausages, nuggets, fish fingers, and processed ham; sodas and other sweetened drinks. Here are many potential health effects of ultra-processed foods:

1. INCREASED CANCER RISK

A five-year study of over 100,000 people found that every 10 percent increase in consumption of ultra-processed food was associated with a 12 percent higher risk of cancer.

2. TOO MUCH SUGAR, SODIUM, AND FAT

Heavily processed foods often include unhealthy levels of added sugar, sodium, and fat. These ingredients make the food we eat taste better, but too much of them leads to serious health issues like obesity, heart disease, high blood pressure, and diabetes.

3. LACKING IN NUTRITIONAL VALUE

Heavy processing strips many foods of their basic nutrients, which is why many foods today are fortified with fibre, vitamins, and minerals.

4. CALORIE-DENSE AND ADDICTING

It's very easy to overindulge in unhealthy food and consume more calories than we realize. For example, an Oreo cookie contains about 50 calories, while an entire cup of green beans is only 44 calories. Processed foods like these are also designed to stimulate our brain's 'feel-good' dopamine centre, making us crave more of them in the future.

5. QUICK TO DIGEST

Processed foods are easier to digest than unprocessed, whole foods. That means our bodies burn less energy (hint: calories) digesting them. It is estimated we burn half as many calories digesting processed foods compared to unprocessed foods. This fact, combined with the calorie density of processed foods in general, can make it easy to pack on the pounds.

6. FULL OF ARTIFICIAL INGREDIENTS

There are about 5,000 substances that get added to our food. Most of them have never been tested by anyone other than the company using them. That includes additives to change colour, texture, flavour, and odour as well as preservatives and sweeteners.

10 TRICKS OF FOOD LABELLING TO KNOW FOR BETTER HEALTH

When grocery shopping, read nutrition labels and avoid foods with high sodium and sugar content. As front labels try to lure you into purchasing products by making health claims, one of the best tips may be to completely ignore such claims. Manufacturers are often dishonest in the way they use these labels. These health claims are mostly misleading and, in some cases, downright false. Thus, a better practice is to study the ingredients list. Remember, product ingredients are listed by quantity – from highest to lowest amount. A good rule of thumb is to scan the first three ingredients, as they make up the largest part of what you are eating. If the first ingredients include refined grains, a type of sugar, or hydrogenated oils, you can assume that the product is unhealthy. Instead, try choosing items that have whole foods listed as the first three ingredients. In addition, an ingredients list that is longer than two to three lines suggests that the product is highly processed. Here are some of the most common claims made by packaged-food industry – and what they mean:

1. NO ADDED SUGAR

Just because sugar has not been added doesn't necessarily mean the food or drink will be low in sugar. Fruit juice is a fantastic example of this. Remember that dextrose, fructose, galactose, glucose, lactose (or milk solids), maltose, and sucrose are some other names by which sugar is hidden in our labels; there are 100 variants to sugar, so do your research before buying anything that claims to be sugar free, unsweetened, or has no added sugar. As the ingredients are listed in descending order by weight, the relative position of sugar – by any of its names – in an ingredients list identifies the amount of sugar the product contains.

2. NATURAL

This does not necessarily mean that the product resembles anything natural. It simply indicates that at one point the manufacturer worked with a natural source like apples or rice.

3. BAKED NOT FRIED

If you are baking something at home, it will very often be healthier if you bake it rather than deep fry. However, that is not necessarily the case when buying packaged food. Quite often, the product will be placed on the tray, sprayed generously with oil (that is, quite often poor-quality vegetable oil or palm oil), then baked, so frying may as well have happened.

4. MADE WITH WHOLE GRAINS OR MULTIGRAIN

The product may contain very little whole grains. Check the ingredients list; if whole grains aren't the first three ingredients, the amount is negligible. In the case of multigrain, it can mean that a product contains more than one type of grain. These are most likely refined grains – unless the product is marked as whole grain.

5. GLUTEN-FREE

This does not necessarily mean healthy. The product simply doesn't contain wheat, spelt, rye, or barley. Many gluten-free foods are highly processed and loaded with unhealthy fats and sugar.

6. CHECK FOR SALT

The nutritional facts label on packaged food lists milligrams of sodium per serving, so it's important to note how many servings the container holds. The percent daily value is based on 2,300 mg, so if your own daily sodium limit is lower, the amount of sodium in a serving is actually a higher percentage than the label indicates. Be aware that some over-the-counter drugs also contain sodium.

7. FRUIT-FLAVOURED

Many processed foods are sold with a name that refers to a natural flavour, such as strawberry yogurt. However, the product may not contain any fruit and instead only chemicals designed to taste like that fruit.

8. CHOLESTEROL-FREE

The liver makes cholesterol, so only animal products like meat, dairy, eggs, and butter can contain it. If a plant-based product (such as sunflower oil) claims to be cholesterol-free, there's no benefit compared to other vegetable oils, which also don't contain it.

9. SUPER FOODS AND SUPER FRUITS

All plant-based foods are packed with vitamins, minerals, and nutrients that are superb for you. There is not one fruit or vegetable that supersedes the others. So don't let these superlatives deceive you.

10. LITE

This refers to a food that has a third of the calories or 50 percent less fat as compared to the original. But it could also mean it is light in colour or flavour, as in 'light' olive oil, or light in cut, as in 'lite' potato chips. Read the nutrition panel to see what's really going on.

OLD POISON, NEW BOTTLE – BIG FOOD, BIG MONEY, AND FAKE FOOD

The industrial West has always been arrogant and ignorant of the cultures it has colonized. 'Fake food' is just the latest step in a history of food imperialism. The corporations and billionaires who have declared war against the Earth and created multiple crises are now preparing for the intensification of industrialized agriculture through digitalization and artificial intelligence (AI). They are envisioning a future of farming without farmers and fake food from labs, which will inevitably deepen the ecological crisis by destroying biodiversity. Drones can spray Roundup (pesticide) but they cannot love the Earth and care for her. Only our heads, hearts, and hands working with the earth, as integral parts of her creativity, can heal it.

The mad rush to fake food, like fake meat made with genetically modified soy, ignores the importance of the diversity of our foods and culinary cultures. It's a recipe to accelerate the destruction of the planet and our health. Through fake food, evolution, biodiversity, and the web of life are being redefined as an 'ancient unimprovable technology', ignorant of the sophisticated knowledge that has evolved in diverse agricultural and food cultures in diverse climates and ecosystems to sustain and renew the biodiversity, the ecosystems, and the health of people and the planet.

Indeed, the promotion of fake foods seems to have more to do with giving new life to the failing GMO agriculture and the junk food industry, and the threat to it from the rising consciousness and awareness that organic, local, fresh food is real food, which regenerates the planet and our health. In consequence, investment in plant-based food companies has soared from near zero in 2009 to $600m by 2018. The AI in the Food & Beverages Market size is estimated at $7 billion in 2023, and is expected to reach $35.42 billion by 2028, growing at a CAGR of 38.30 percent during the forecast period (2023–28).

Ecological sciences have been based on the recognition of the interconnections and interrelatedness between humans and Nature, between diverse organisms, and within all living systems, including the human body. It has thus evolved as an ecological and systems science, not a fragmented and reductionist one. Diets have evolved according to climates and the local biodiversity the climate allows. The biodiversity of the soil, of the plants, and our gut microbiome is one continuum. In Indian civilization, technologies are tools. Tools need to be assessed on ethical, social, and ecological criteria. Tools and technologies have never been viewed as self-referential. They have been assessed in the context of contributing to the well-being of all.

The growing artificial food industry, with a thrust on plant-based alternatives, is a grave health challenge. An example of this is the Impossible Burger, a plant-based alternative to traditional meat-based burgers. It's said to mimic the flavour, aroma, and texture of beef. In an article, 'How our commitment to consumers and our planet led us to use GM soy', Pat Brown, CEO & Founder of Impossible Foods states that, 'We sought the safest and most environmentally responsible option that would allow us to scale our production and provide the Impossible Burger to consumers at a reasonable cost.' Given the fact that 90 percent of the monarch butterflies have disappeared due to Roundup Ready crops, and we are living through what scientists have called an 'Insectageddon', using GM soya is hardly an 'environmentally responsible option'. In writing this, Pat Brown revealed his total ignorance that weeds have evolved resistance to Roundup and have become 'super weeds', now requiring more and more lethal herbicides. At a time when across the world the movement to ban GMOs and Roundup is growing, promoting GM soya as 'fake meat' is misleading the consumer both in terms of the ontology of the burger and on claims of safety.

The 'Impossible Burger', based on GMO, Roundup-sprayed soya, is not a 'safe' option. The levels of glyphosate detected in the

Impossible Burger by Health Research Institute Laboratories were 11 times higher than the Beyond Meat Burger. The total result (glyphosate and its breakdown AMPA) was 11.3 ppb. Moms Across America also tested the Beyond Meat Burger, and the results were 1 ppb. 'We are shocked to find that the Impossible Burger can have up to 11x higher levels of glyphosate residues than the Beyond Meat Burger, according to these samples tested. This new product is being marketed as a solution for "healthy" eating, when in fact 11 ppb of glyphosate herbicide consumption can be highly dangerous. Only 0.1 ppb of glyphosate has been shown to destroy gut bacteria, which is where the stronghold of the immune system lies. I am gravely concerned that consumers are being misled to believe the Impossible Burger is healthy.'

---

1. *Ultra-processed food intake and risk of cardiovascular disease: prospective cohort study (NutriNet-Santé) The BMJ 2019, 365. https://doi.org/10.1136/bmj.l1451 (Published 29 May 2019)*

SARSON SATYAGRAHA

Women from India's slums reached out to Navdanya to bring our mustard back when GM soya oil started to be dumped on India, and local oils and cold-press units in villages were made illegal. That is when we started the 'Sarson (mustard) Satyagraha' to defend our healthy cold-pressed oils from the dumping of hexane-extracted GM soya oil. Hexane is a neurotoxin.

While Indian peasants knew that pulses fix nitrogen, the West was industrializing agriculture based on synthetic nitrogen, which contributes to greenhouse gases, dead zones in the ocean, and dead soils. While we ate a diversity of 'dals' in our daily 'dal roti', the British colonizers, who had no idea of the richness of the nutrition of pulses reduced them to animal food. Chana became chickpea, gahat became horse gram, and tur became pigeon pea.

We stand at the precipice of a planetary emergency, a health emergency, and a crisis of farmers' livelihoods. Fake food will accelerate the rush to collapse. Real food gives us a chance to rejuvenate the earth, our food economies, food sovereignty, and food cultures. Through real food, we can decolonize our food cultures and our consciousness. We can remember that food is living and gives us life.

Reclaiming, Understanding Holistic Healing and Traditional Medicine

'Medicine's a funny business. After all, dispensing chemicals is considered mainstream and diet and nutrition is considered alternative.'
CHARLES F. GLASSMAN
Brain Drain: The Breakthrough That Will Change Your Life

Across the world, there is an intense contest emerging between two paradigms of health and science. The first paradigm is holistic and sees connections between the health of the planet and our health. It is based on the ecological science of interconnectedness. The second paradigm is reductionist, mechanistic, and commercial, wherein the mechanistic worldview sees us as separate from nature and each part of our body as separate from all others, as parts of a machine are. Health is defined as a commodity we buy from the pharmaceutical industry.

The mechanistic paradigm has transformed the diversity of knowledge systems into a hierarchy, privileging the mechanistic and reductionistic paradigm as the only science and pushing all other knowledge systems to oblivion or treating them as inferior.

In the early 1980s, a group of companies successfully put intellectual property at the top of the US trade agenda and helped to design not only international agreements but also a very successful strategy for convincing other countries to sign onto them. Prominent in that group of companies were representatives of the pharmaceutical industry: the Advisory Committee for Trade Negotiations, formed in 1981, was chaired by the CEO of Pfizer, Ed Pratt, and its 1986 successor, the Intellectual Property Committee, consisted of thirteen large companies, including Pfizer, Bristol-Myers, Johnson & Johnson, and Merck, thereby making pharmaceuticals the largest of the industry sectors represented on the committee. The result was the 1995 Trade-Related Aspects of Intellectual Property Rights (TRIPS) Agreement, which has formed the basis for new intellectual property laws in countries around the world. Effectively, TRIPS extended a version of a new and strengthened US intellectual property law to many parts of the globe.

TRIPS and 'access to medicines' were so strongly linked that in the early 2000s almost all of the most prominent discussions of access also referred to it or to patent protection. The most immediate concerns that linked the two were about antiretroviral treatments, given the HIV/AIDS crises in sub-Saharan Africa, Southeast Asia, and Brazil. While on one hand access to medicine becomes difficult with rising drug prices due to these World Trade Organization laws, on the other hand, it opens up possibilities of intellectual robbery of indigenous medicinal knowledge systems by these companies.

Plants are invaluable, incredible, and traditional sources for the curability of various diseases in the form of medicines. According to the World Health Organization (WHO), out of the 252 drugs considered as basic and essential, 11 percent are plant-based. WHO also says that about three-quarters of the world population depends on traditional remedies. Plants being used as medicine are not a recent phenomenon, but their use as traditional and folklore drugs throughout the world for health services is a matter of tremendous historical legacy and evidence. Consequently, herbal medicines have been used for centuries for safety, efficacy, fewer side effects, and cultural acceptability.

Pharmaceutical companies make billions from addictive painkillers. They don't want you to know about safer alternatives offered by holistic healthcare practitioners. Subsequently, now they are patenting traditional medical knowledge by misappropriation by unauthorized third parties, who have patented compounds derived from traditional medicines without the prior consent of traditional medical knowledge holders and without fair compensation. Examples of patents based on traditional Indian medicine have included the use of turmeric for healing wounds, the anti-fungal properties of neem, and a diabetes medicine made from the extract of jamun. All three patents were subsequently revoked. We will discuss more about these cases in the chapter on Slow Knowledge.

Many indigenous systems such as Ayurveda, yoga, Unani, homeopathy, naturopathy, and Siddha are famous and have been followed in India for decades. From ancient times, the treatment of many diseases was done by modulating the immune system or function by using medicinal plants and their products; this was also the fundamental principle of the therapeutic approach.

3 TRADITIONAL HEALING PRACTICES TO EXPLORE FOR BETTER HEALTH

1. FASTING AS A SUPER MEDICINE

In 2016, the Nobel Prize for Medicine was awarded to a Japanese scientist Dr Yoshinori Ohsumi for his research on autophagy. Autophagy, or 'self-eat', refers to the process by which the human body eats its own damaged cells and unused proteins. It is a natural process, one which occurs in cases of starvation. The failure of autophagy is one of the main reasons for the accumulation of damaged cells, which eventually leads to various diseases in the body. Autophagy is important to prevent and fight cancer, playing a vital role in degrading and 'consuming' cells infected by bacteria and viruses. Fasting is part of most spiritual traditions in the world. In ancient India, it was a recommended practice to fast one day in a fortnight ('Ekadasi'). Through this process of fasting-induced autophagy, our body repairs its damaged and degenerated cells and also uses up the damaged cells' proteins. The current trend of intermittent fasting can also be linked to the common Indian adage where people said 'do waqt ki roti' for employment – which literally translates to 'bread twice a day', suggesting that eating three to four meals a day was not the tradition earlier. Clearly, our ancients believed in a process of holistic healing of both the body and the mind.

2. SLEEP HYGIENE

According to most healing traditions, the quality and quantity of our sleep are as essential to our health and well-being as our dietary habits. Ayurveda teaches that there is a certain period of the day that is optimal for sleeping, and that the amount of sleep we need varies individually (and even seasonally). Lack of adequate sleep interferes with the ability of our bodily systems to repair and rejuvenate. Without sufficient rest, our digestion, metabolism, cell regeneration, emotional balance, creativity, mental clarity, and motor skills are impaired. There are a number of factors that can inhibit sound sleep, including lifestyle, environmental stimuli, travel, and health disorders, along with our emotional and mental state. In our modern '24-hour world', where we can work and play whenever we choose, we are still intrinsically regulated by the cycles of nature. Ayurveda proposes that the more we align ourselves with nature's rhythms, in accordance with our individual nature, the more we will experience a life in balance. The current sleep aid industry, according to a research market, was projected to reach up to $114.15 billion in 2025, with a compound annual growth rate of 7 percent over the forecast period (2019–2025). Electronic devices emit an artificial blue light that can suppress the release of the body's sleep-inducing hormone, melatonin. In turn, this can interfere with the body's natural internal clock. Limiting overall daily screen time can help improve sleep issues, but most importantly, restricting the use of any gadgets right before bedtime can play a key role in better sleep hygiene. Therefore, listening to our body clock, understanding our personal sleep patterns, and being aware of the food that excites us as well as the food that relaxes us are a must for better sleep hygiene.

According to Ayurveda, here are a few recommendations that promote sound sleep:

• Follow a regular routine of going to bed in the kapha time of the evening, between 8 to 10 pm, and waking up in the vata time of the morning before sunrise.

• Create a pre-bedtime routine – such as abhyanga (oiling your body) or soaking your feet in Himalayan salt, calming yoga asanas, pranayama (breathing exercise), visualization, meditation, drinking a cup of warm milk or calming herbal tea, listening to soothing music or sounds of nature. See what works for you, then make it into a routine.

• Create a bedroom that is clean, uncluttered, and peaceful. Use soft colours and textures. Avoid watching TV, using a computer, talking on the telephone, or eating in this room.

• Schedule your meals. Make your last meal of the day easily digestible, light, and nourishing. Allow yourself two to three hours after eating and before going to bed.

• Avoid naps late in the afternoon, such as after 3 pm. Exercise regularly. However, do not exercise too late in the day, as undertaking such activities close to bedtime may make it difficult to fall asleep.

3. MINDFULNESS

There are several spiritual and non-religious practices for you to explore the techniques of mindfulness. To practice mindfulness usually involves breathing methods, guided imagery, and other methods to relax the body and mind. Almost all religions have prayer beads for chanting, and even Western psychology has grounding techniques to deal with anxiety and visualization exercises for dealing with different kinds of stress. Practising mindfulness can give more insight into emotions, boost one's concentration, and improve relationships. Here are some simple steps to get you started on the journey:

• **Set aside time**. You don't need a meditation cushion, bench, or prayer bells to access your mindfulness skills, but you do need to set aside some time and space. Choose a specific time in the day that works for you and choose a time limit; even ten minutes is good to begin with.

• **Observe the present moment as it is**. The aim of mindfulness is not quieting the mind or attempting to achieve a state of eternal calm or 'nirvana', but to pay attention to the present moment without any judgement. To sit in awareness and stillness without getting annoyed with oneself when the mind wanders off, but bringing it back to the present moment is essential. You can do this while cutting vegetables, cooking, walking, or even sitting. The point is to be completely present in the moment.

• **Let your judgments roll by**. When you notice judgments arise during your practice, make a mental note of them and let them pass.

• **Be kind to your monkey mind**. Don't judge yourself for whatever thoughts crop up. Just practise recognizing when your mind has wandered off and gently bring it back.

GARDEN OF HOPE – HARVEST OF HOPE

In contrast to the busy harvest time at a farm, harvesting vegetables in your garden can be quite a fun experience. After great patience and care, the feeling of holding fresh vegetables that are a result of your own effort is most rewarding. While cooking and eating them is a whole-other-mouthful, beyond-words experience.

Different vegetables have varied harvest periods. The simplest way for you to judge is to compare your plant with the same you would purchase from a market. But here are some pointers that will give you an idea about when and how to harvest the plants you would have grown:

- Cauliflowers take about three to three-and-a-half months to mature. The vegetable may vary in shape, size, and colour, but the lack of uniformity does not indicate any shortcomings in taste. You may consider the cauliflower ready when it is hard and doesn't press easily. They need to be cut with a sickle about an inch or two from the ground.

- Mustard, too, takes the same amount of time. However, after a month, leaves that grow may be harvested once, then given time to grow again and harvested a second time for good use. Thereafter, the plant should be permitted to grow its full three to three-and-a-half month span until it flowers and provides seeds.

Carrots and radishes take about two months. As they are root vegetables, an indication to know they are ready is when one or two carrots/radishes start pushing their tops out of the ground. One can wiggle them out of the soil rather than pull forcefully, as they break easily.

Spinach and leafy greens grow relatively fast. The initial growth of the sprout requires patient waiting, but once that is done, it grows fast. Their leaves can be harvested for the first time a month and a half or so after sowing. Thereon, their leaves can be allowed to re-grow and harvested about three times before the plant exhausts its capacity to give.

Coriander and fenugreek take about a month-and-a-half to two months to be ready for consumption. While coriander takes a longer time to sprout after sowing, it grows quickly once it has taken root; fenugreek sprouts quickly, but takes a while to grow into its mature edible form. Both will be ready around the same time. They also give at least three to four rounds of produce before living out their span. To have these greens regrow their leaves for use, we should be careful not to damage the roots or stems while collecting the leaves.

If one wants to save seeds for the next season, the plants should not be harvested. A few healthy plants should be allowed to live out their entire lifespan. The plant will thereafter bud, flower, and give seeds. These seeds should be collected and stored in airtight containers.

After harvesting, the vegetables need to be washed under running water before being consumed. They can be eaten raw for a healthier diet or can be pickled/cooked in mouthwatering ways.

READING LIST

GABOR MATÉ, *When the Body Says No: Exploring the Stress-disease Connection. Hoboken.* N.J., J. Wiley, 2011.

RAY MOYNIHAN AND ALAN CASSELS, *Selling Sickness: How the World's Biggest Pharmaceutical Companies Are Turning Us All Into Patients.* Nation Books, 2005.

MICHAEL POLLAN, *In Defense of Food.* Penguin, 2009.

RITU PRIYA AND SYEDA HAMEED, *Dreams of a Healthy India: Democratic Healthcare in Post-Covid Times (Rethinking India Vol. 9)*, 2023.

Recipes to Boost Immunity: Grandmothers' University Special

At Navdanya we celebrate the wisdom of women's health and healing, and we pay homage to the wisdom of the traditional kitchens and our grandmothers who have healed us with their food and have been the keepers of our traditional wisdom. Bija Vidyapeeth, or our Earth University in Dehradun, has had the privilege of hosting and sharing this wisdom through our Grandmothers' University courses. Here are some jewels shared across the years to enrich your kitchens and celebrate the healing power of food and drinks.

TULSI KADHA

INGREDIENTS

1 litre water
2 tsp black peppercorns, coarsely ground
2 tsp cloves, coarsely ground
2-inch piece cinnamon
4–5 green cardamom pods, lightly pounded in a mortar & pestle
1 tbsp ginger, pounded in a mortar & pestle
1 fistful holy basil leaves, shredded
2 tbsp jaggery (gur), or as per taste
1 tbsp honey, or as per taste

METHOD

- In a heavy-based saucepan over high heat, transfer the water. Add all the kadha ingredients – black peppercorns, cloves, cinnamon, green cardamom, ginger, holy basil leaves, and jaggery, and bring to a boil.
- Partially cover the pan and let simmer until the water reduces to half the quantity, about 500 ml. Remove from the heat and add honey.

An aromatic herbal brew and an immunity-boosting remedy for sore throat, cough, cold, and seasonal infections.

IMMUNITY-BOOSTING **TURMERIC PICKLE**

INGREDIENTS

100 gm raw turmeric
50 gm ginger
6 green chillies
6 garlic pods
½ cup lemon juice
1½ tsp salt

SPICE POWDER MIX

3 tsp mustard seeds
1 tsp fennel seeds
¼ tsp fenugreek seeds
¼ tsp asafoetida

OTHER INGREDIENTS

2 tbsp mustard oil

Curcumin, the active ingredient of turmeric, has anti-inflammatory, antioxidant, anti-bacterial, anti-viral, and anti-microbial properties.

METHOD

- Peel the skin of turmeric root and ginger. Cut them into thin slices. Peel the garlic and slice it into thin pieces.
- Transfer the turmeric, ginger, green chillies, and garlic into a glass bowl. Add in half cup lemon juice and salt as required. Mix well and set aside.
- Into the mixer jar, add the mustard seeds, fennel seeds, fenugreek seeds, and asafoetida. Grind to a coarse powder and set aside.
- Mixing the Pickle: Heat the mustard oil in a wide pan. Once the oil is hot, add the prepared spice powder mix. Turn off the heat immediately and give it a nice mix.
- Now add the turmeric and other ingredients. Mix well so that the spices are nicely coated. Cover with a lid and set aside for a few hours.
- Give it a nice stir after a few hours and adjust salt, if needed.
- Transfer the pickle into a clean and dry glass jar. You can leave it for a day or two at room temperature. Stir the pickle now and then so that all the juices released are mixed evenly.
- Refrigerate later for longer use. You can use it for about 3 to 4 weeks.

HOME-MADE HEALTHY **CHYAWANPRASH**

INGREDIENTS

- ½ kg Indian gooseberry/amla
- 400 gm jaggery
- ⅓ cup ghee

TO BE POWDERED

- Few saffron strands
- 1 bay leaf
- 1 inch cinnamon stick
- 10 gm dry ginger
- 10 gm long pepper
- 5 gm nutmeg
- 5–7 pieces of green cardamom
- 5 gm clove
- 5 gm black pepper

Made with Indian gooseberry (amla) as the main ingredient, Chyawanprash has 40 other herbs and spices making this a potent immunity booster. Here is a simple version for you to try at home.

METHOD

• Wash the amla thoroughly. Take two cups of water in a pressure cooker. Add the amla and pressure cook for two whistles. Meanwhile, break the jaggery to very small chunks and keep aside.

• Take all the ingredients under the list 'to be powdered' into a mixer-grinder, and grind to a very fine powder. Sieve the powdered spices and herbs.

• Once the pressure releases on its own, remove the amla. Allow them to cool a bit and de-seed them. Put the amla along with a little water used for cooking in a blender and blend to a smooth, lump-free puree.

• Heat a wide-bottomed wok over low heat and add the ghee. Once the ghee melts, add half the spice blend. Now add the amla puree. Stir well and cook over low-to-medium heat until the liquid evaporates and the mixture reaches a thick consistency. This should take 5–7 minutes.

• Now add the jaggery and stir well. Keep stirring the mix until the jaggery melts and it begins to thicken again, and the jam starts looking dark brown.

• Once the jam thickens, add the other half of the spice blend and keep stirring.

Source: Masale – Spices, *Navdanya*

6

SLOW CLOTHING

The Charkha Way

'Becoming more mindful about clothing means looking at every fibre, at every seed and every dye and seeing how to make it better. We don't want sustainability to be our edge, we want it to be universal.'
EILEEN FISHER
Sustainable Fashion Designer

The British Empire was an empire of cotton. When we were free, a sack of indigo could be exchanged for a sack of gold. As a British colony, peasants in Bengal and Bihar were forced to grow indigo under slave conditions. Movements against forced indigo cultivation were growing, with people like Rajendra Prasad, who later became India's first president after Independence, and other freedom fighters supporting the peasants. 'We would rather die than grow Indigo,' was the cry of the peasants. Gandhi's first Satyagraha after he returned to India from South Africa in 1915 was the Indigo Satyagraha in Champaran (Bihar) in 1917.

Apart from destroying India's textiles, the British Empire colonized the land of the indigenous peoples of the Americas, captured Africans and made them slaves on the American cotton plantations. This cotton was taken to the mills of Lancashire and Manchester, and the industrial textiles were dumped on the world. Gandhi took out the spinning wheel as an instrument of freedom. The spinning wheel is the symbol of Slow Clothing. We call it the charkha way.

Today, fast fashion is destroying slow clothing, local economies, and local cultures. It is destroying the biodiversity of fibres and cotton. Almost 95 percent of the cotton is now GM Bt cotton, which has trapped farmers in debt and pushed bees and pollinators to extinction. More than 80 percent of the farmers' suicides in India since 1995 are in the cotton belt.

I was born soon after Independence. We had charkhas at home, and wearing hand-spun, hand-woven khadi was the norm. Then fossil fuels were turned into nylon, and petrochemicals started to shape clothing. 'Glass nylon' entered India as the new fashion. I remember my mother asking my sister and me what we wanted for our birthdays. Everyone around us was wearing nylon, so we said 'a nylon frock'. Our mother replied that she would get it for us, but we should be aware that when you wear nylon, a rich industrialist gets his next Mercedes car. When you wear khadi, a poor woman could feed her child that night. You make the choice. And since that day, the choice I have been making is the choice from consciousness: Does what I wear leave the Earth and fellow human beings in a better condition because of my choice, or does it leave the planet and people poorer, more polluted, and more exploited? This consciousness shapes our ideas of beauty. For me, handmade is beautiful. I look at junk clothing as junk – embodying exploitation of the Earth, of people.

Slow clothing is wearing clothes as an expression of mindfulness of our relationship with all beings, an expression of our culture, and an expression of who we are. As an expression of place and culture, clothing was made of diverse fibres, not just cotton. In cold climates, wool obtained from different animals provides clothing – camels, sheep, yak, and llama. To know the natural fibres from the synthetic ones and natural dyes from synthetic dyes, along with the impact the production of our clothes has on the planet and the people making it, are all part of the simple act of wearing clothes in slow fashion.

A study conducted in 2015 shows that 97 percent of what goes into making clothes are new resources, with only 3 percent of it being recycled materials. This adds up to an annual resource input of 98 million tonnes – including oil to produce synthetic fibres, fertilizers to grow cotton, and an endless list of chemicals needed to dye and finish fabric. Thus, your wardrobe is a statement on what kind of planet you want to live in. Fashion reflects the quality of heart and temperance of your mind: Your outer beauty reflects your inner choices. If we look at the act of dressing up each day as an act of personal and social transformation, the quality and quantity of what we buy and put on will dramatically change. As we do this, we will also transform the markets that depend on our purchases and need us to consume their products. So, as you read this chapter, one hopes that you will open your cupboards and reflect with us on what kind of world lives inside them.

Fast Fashion vs. Slow Clothing

'There is no beauty in the finest cloth if it makes hunger and unhappiness.'
MAHATMA GANDHI

Before the 1800s, fashion was slow. People had to source their own materials like wool or leather, prepare them, weave them and then make the clothes. After the Industrial Revolution, when new technology was first introduced – such as the sewing machine – clothes became easier, quicker, and cheaper to make. Dressmaking shops emerged to cater to the middle classes. A lot of these dressmaking shops used teams of garment workers or home workers. It was around this time that sweatshops emerged, along with some familiar safety issues. The first major garment factory disaster was when a fire broke out in New York's Triangle Shirtwaist Factory in 1911. It claimed the lives of 146 workers, many of whom were young, female immigrants. A hundred years later, in November 2012, a fire in the Tazreen Fashions factory in Bangladesh killed at least 112 workers – indicating nothing much had changed in the fashion industry and, in fact, the conditions had probably worsened.

In the 1960s and '70s, new trends were emerging and clothing became a form of personal expression, but there was still a distinction between high fashion and high street. In the late 1990s and 2000s, low-cost fashion reached its zenith. Online shopping took off. Fast fashion retailers like H&M, Zara, and Topshop took over the high street. These brands took the looks and design elements from the top fashion houses and reproduced cheap variants in no time. With everyone now able to shop for on-trend clothes whenever they want, it's easy to understand how the phenomenon caught on. Globally, clothes consumption is predicted to rise by 63 percent by 2030. This means an increase from 62 million metric tonnes (in 2017) to 102 million metric tonnes in 2030 – or more than 500 billion T-shirts.

FAST FASHION	SLOW CLOTHING
Factory-made and mass production – indifference to conditions of those working on farms and in factories with an unsafe work environment and low wages.	Handloom and artisanal – where labour is not a cog in the wheel but an artist, and every piece made with its imperfection and beauty is the expression of their creativity.
Monoculture – a thoughtless clone of millions of others, wearing the same throwaway clothing, of the same colour, irrespective of the culture and climate to which you belong. Indifference to your unique identity.	Unique – no two handloom products are the same, as they are handmade and may have irregularities that add to their beauty.

Indifference to the ecological costs to the Earth. The quick response model and new supply chain practices of fast fashion even accelerate the speed of it. As a result, excessive stock and untrendy clothes tend to end up in landfills.	Eco-friendly – the means of production are not chemically driven or water-intensive, and the raw material is biodegradable.
Use and throw – In recent years, the fashion cycle has steadily decreased as fast fashion retailers sell clothing that is expected to be disposed of after being worn only a few times. This dramatically shortens the consumers' buying cycle. The quick-changing stocks and low prices of fashion goods encourage consumers to visit the store and buy more frequently.	Cherish and mend – each piece is unique and a piece of art, and the philosophy is to consciously buy things that you love and will cherish and mend if torn.
Plastic and toxic fibres.	Natural fibres.
Dy(e)ing planet/synthetic dyes.	Colour of life/natural dye.

10 FACTS ABOUT FAST FASHION THAT YOU NEED TO KNOW

1. According to Fashion United, the global apparel market is valued at 3 trillion dollars (3,000 billion dollars) and accounts for 2 percent of the world's Gross Domestic Product (GDP).

2. The 2015 documentary *The True Cost: The Bitter Truth Behind Fast Fashion* shows that '80 billion pieces of clothing are consumed globally every year'.

3. According to the Ellen MacArthur Foundation, 'Global clothing production has doubled in the past 15 years, with garments on average being worn much less and discarded quicker than ever before.'

4. In an article by James Conca, published in *Forbes* Magazine (2015), he elucidated the cost of fast fashion and claimed, 'Nearly 70 million barrels of oil are used each year to make the world's polyester fibre, which is now the most commonly used fibre in our clothing. But it takes more than 200 years to decompose.'

5. According to WWF, '20,000 litres is the amount of water needed to produce one kilogram of cotton; equivalent to a single T-shirt and pair of jeans.'

6. *The True Cost* shows that 'more than 90 percent of cotton is now genetically modified, using vast amounts of water as well as chemicals. Cotton production is now responsible for 18 percent of worldwide pesticide use and 25 percent of total insecticide use.'

7. According to the ecologist Dr Mark Browne, '85 percent of the plastic pollution in the ocean is due to microfibres from synthetic clothing.'

8. According to an article published in *Forbes*, 'Fast fashion garments, which we wear less than five times and keep for 35 days, produce over 400 percent more carbon emissions per item per year than garments worn 50 times and kept for a full year.'

9. According to the website Common Objective, 35 percent of materials in garment supply chains are wasted. For every 100,000 people, 100 kg of microfibres are released through washing clothes. Almost 57 percent of all discarded clothing ends up in a landfill.

10. According to Oxfam Australia's report 'What She Makes', 'women aged 18–25 make up 80 percent of the factory workers in the global garment industry'. In Bangladesh, the local minimum wage equates to just 39 Australian cents an hour. In Vietnam, it is just 64 cents, and in China, it is 93 cents. Based on the CEOs' pay levels of some of the big brands in Australia, it would take a Bangladeshi garment worker earning the minimum wage more than 4,000 years to earn the same amount that a CEO gets paid in just one year.

Detox Your Wardrobe for a Plastic-Free Future

'Everything we make returns to the Earth either as food or as poison.'
CELINE SEMAAN
Founder, Slow Factory

See-through backpacks, flip-flops, and promotional sunglasses are obviously made of plastic. But do you know that 64 percent of all clothing worldwide is also made from plastic fabrics? Polyester, nylon, and almost all vegan leather, and also any other type of man-made material moulded from petroleum are all plastic. Apparel brands are legally required to name the composition of fabrics and materials in their products, and it's often a blend of two or more synthetic and natural fibres. So every time we wash these materials, they shed millions of plastic microfibres. Threads so small they can drain out of our washing machines and pass straight through wastewater treatment plants into the sea. They might be minuscule, but these plastic pieces can absorb high concentrations of poisonous substances, including chemicals we once used in products like pesticides, which are now banned, like DDT. Sea creatures, especially plankton, absorb these toxic microfibres, and they pass them up the food chain. Plastic-contaminated fish, mussels, and other seafood that we eat have a high probability to make it to our plates and into our stomachs.

Some brands market polyester apparel as environmentally friendly because they use recycled plastic bottles to make it. That may sound like a good way to reduce plastic pollution, but these garments shed lots of plastic debris. For many materials, recycling is a useful way of preventing pollution – but not for plastic. It just delays the inevitable escape of pollutants into the environment. And many plastics can only be recycled a few times before they become too low-grade anyway. That's why we need to phase out all but the most essential plastics. We certainly don't need them in our clothes and textiles. To start you on the journey of detoxing your wardrobe from plastic, here is a list of plastic-based fibres that you should avoid buying:

1. POLYESTER

Accounting for more than half of global consumption, polyester is used the most by the fashion industry, as it's cheaper than natural fibres, easily produced, durable, and has technical properties that natural fibres don't have, such as being waterproof, stretchable, and/or wrinkle-free.

2. NYLON

This synthetic fibre is primarily used in stockings and some technical gear.

3. ACRYLIC

This is the least popular synthetic fibre, though still more popular than what it replaces: wool. You will find this in cheap sweaters and some faux fur products.

4. POLYVINYLCHLORIDE AKA PVC AKA VINYL

An extremely toxic material to produce, PVC also continues to emit toxic fumes once it is in your home or on your body. Mostly used in shower curtains, this product has also become popular in recent years in the form of vegan leather.

5. POLYURETHANE AKA PU

While not as toxic to produce or own as PVC, it is still a kind of synthetic plastic that replaces leather in fashion accessories.

6. ELASTANE AKA SPANDEX AKA LYCRA

A synthetic fibre added to natural fibres to give them stretch. You will find this in underwear, bathing suits, and bras.

7. OTHER NAMES FOR PLASTIC IN YOUR CLOTHES

Spectra, Dyneema, Econyl, Odegon, and Zylon.

8. Basically, anything that is static-resistant, stain-resistant, permanent press, wrinkle-free, stain-proof, or moth-repellent. Many of the stain-resistant and wrinkle-free fabrics are treated with perfluorinated chemicals (PFCs), like Teflon.

So, what do these synthetic fibres do to our bodies? Wearing polyester or spandex makes it hard for your skin to breathe. Moreover, rising body temperatures help to release chemicals from polyester that are later absorbed by your skin. This can cause a variety of skin problems and irritations like rashes, itching, redness, eczema, and dermatitis. So it's better to avoid polyester in your clothes and also in your linens. Rayon fabric can emit toxic substances that can cause nausea, headache, vomiting, chest and muscle pain, and insomnia. Nylon clothes do not absorb sweat from the skin, which can cause bad odours and skin infections. While in its production, the fabric gets bleached or dyed with different chemicals, wearing it on your skin can cause a variety of irritations as well. Acrylic fabrics are made of acrylonitrile, which is a carcinogen and a mutagen. Exposure to this substance could lead to headache, nausea, dizziness, difficulty breathing, limb weakness, and more. By wearing acrylic fabrics, you run the risk of absorbing some of the acrylonitrile into your skin.

If we have inspired you to detox your wardrobe, let us now look at the alternatives. Here is a list of plant-based fibres (cellulosic fibres) that you can consider to swap the plastic in your wardrobe with:

FIBRE	SOURCE	ATTRIBUTE
Abacá	Abaca plant	Thin, lightweight
Bamboo	Grass pulp	Lightweight, pliable fibre
Banana	Banana plant pseudostem/leaves	Warm, thick, durable
Kapok	Pentandra tree	Fluffy
Coir	Coconut	Strength, durability
Cotton	Shrub	Lightweight, absorbent
Flax	Herbaceous plant	Lightweight, absorbent, used to make linen
Hemp	Cannabis	Strength, durability
Jute	Vegetable plant in the linden family	Strength, durability
Kenaf	*Hibiscus cannabinus*	Rough

Lyocell	Eucalyptus tree	Soft, lightweight, absorbent
Modal	Beech tree	Soft, lightweight
Piña	Pineapple leaf	Soft, lightweight
Raffia	Raffia palm	Carpet/rough
Ramie	Flowering plant in the nettle family	Heavy, tough
Rayon	Wood pulp; though manmade, rayon is cellulosic	Soft, lightweight, absorbent
Sisal	*Agave sisalana*	Strength, durability

Some other plant-based fibres are *Apocynum cannabinum*, bast fibre, cedar bark textile, esparto, fique, papyrus, straw, and Tapa cloth.

As you embark on this journey of wardrobe detox, it is important to understand what positive aspects of synthetic fibres you might have to give up so you can work around these challenges. Here are the things that make it difficult to get free of synthetic fibres:

1. STRETCH

This is almost always achieved with synthetics. Yes, one can get a limited amount of stretch through knitwear whose structure allows for stretch in some directions. However, knit fabrics won't yield the same performance qualities – lightweight, tight, slimming, opaque – of super-stretchy polyester or nylon fabric. Often, you will have to be fine with even the most sustainable underwear brands having 5 to 8 percent synthetic fibre mixed in with silk and organic cotton.

2. ALL THAT GLITTERS

Most glitter, being made from etched aluminium bonded to polyethylene terephthalate (PET), is a form of microplastic. Mica is touted as an alternative to plastic-based glitter. The main source of the world's mica is from illegal Indian mines. Instead, use natural *zari* or *gota* to add that bling to your wardrobe.

3. WATERPROOF

The technology for eco-friendly waterproofing is still being developed, so we may have to wait for a while. However, if we need it, recycled or second-hand is our best bet.

4. WRINKLE-FREE

Definitely means plastic is involved, so avoid; always carry an iron instead.

5. ULTRA-PACKABLE, LIGHTWEIGHT

This is also a red flag; check the tag for the composition of material.

Now that you know the different fibres and what you should avoid and what you should buy, what are you to do with the plastic in your cupboard that you had already bought? Here are some simple tips for upcycling and disposing this toxic stock that was in your cupboard:

- Getting rid of half your wardrobe isn't environmentally friendly at all. It's wasteful and merely serves to add to the flood of clothing getting dumped on your local resale shops. So use what you have, repurpose them into things that don't need to be washed as regularly, like bags, footstools, or cushion covers.

- Buy bathing suits and workout clothes made of recycled material, since natural fibres cannot make bathing suits yet.

- High-performance rain jackets, winter puffy coats, and lightweight camping gear will always come in synthetics. The alternative is traditional, heavy, low-performance waxed canvas and shearling. Buying second-hand gear may help reduce the impact of the use of synthetic materials in the wardrobe.

• Let the brands that use synthetic fibres know that they are about to lose your business if they don't have natural fibre options. This will take some time for them to register and go through the process of changing, but every customer who tells them that they would prefer clothing in natural or recycled fibres helps convince them to make the switch.

GM Cotton – A Fashion Disaster

'The empire of cotton has continued to facilitate a giant race to the bottom, limited only by the spatial constraints of the planet.'
SVEN BECKERT
Empire of Cotton: A Global History

Although cotton is a naturally occurring fibre, its production is nonetheless haunted by claims of pollution, exploitation, and slavery. Cotton trades were the fulcrum of colonial empires that were built on the backbones of exploitation, slavery, and mass consumption from the Americas to India. It is no surprise that it remains a means for the new wave of colonization, with the fashion industry monoculture and mass production at the helm of it.

Cotton earned the title 'dirtiest crop' because it's sprayed with some of the worst pesticides, including Bayer's aldicarb, a pesticide that was banned in the USA in 2010 but reapproved by the U.S. Environmental Protection Agency in 2016; Syngenta's paraquat, a highly toxic pesticide banned in the European Union but not in the USA; and Monsanto's glyphosate, classified by the WHO as a 'probable' human carcinogen. In traditional farming of cotton, manual de-weeding practices and growing a crop of cowpea between two rows of cotton was used to suppress the early emerging weeds.

About 100 million households are engaged in growing and producing cotton, and 300 million people work in the cotton sector as a whole. The majority of cotton farmers and workers live in developing countries and work extremely long hours. They are exposed to poisonous substances daily and earn very little in wages. In fact, many of them have unsustainable debts because they are unable to keep up with the employers' demands. Other factors, such as climate change, decreasing prices of cotton, and tough competition from farmers in rich countries, don't make it any easier. Sadly, suicide rates among cotton farmers have been high in the last 20 years. In 2013 alone, 11,772 farmers committed suicide in India, which comes to an astounding figure of 44 farmers a day! A rule of thumb: If a cotton product is outrageously cheap, then know that someone else has paid the price. Add to this mix the emergence of GM Bt cotton, and we have a fashion disaster creating a human and environmental catastrophe, which is unfolding as we speak.

According to its proponents, the benefits of genetic engineering in the agriculture sector is great. GMO crops have been hailed as a way to increase yields by protecting against pests, drought, and disease. However, as Prof. Glenn Stone from Washington University states, 'The first long-term study of Bt adoption and production trends at both national and state levels shows that Bt seeds were not the cause of rising yields and that the insecticide reductions were fleeting.'

GMO cotton is grown on 70 million acres across the world, including the overwhelming majority of cotton grown in the USA, India, Pakistan, and China. While occupying a relatively small percentage of arable land globally (2.4 percent), GMO/chemical cotton crops account for a staggering 25 percent of global insecticide sales. Several pesticides used on cotton are known carcinogens, including Roundup. Not only do these pesticides linger on the clothes worn next to human skin, but the fish, marine, and wildlife surrounding or downstream from cotton fields also suffer from pesticide pollution. Non-organic cotton crops utilize large amounts of chemical fertilizers that routinely pollute groundwater and emit nitrous oxide, one of the most polluting gases.

Most of the world's highly contaminated cotton seeds and cotton gin trash end up in animal feed (especially non-organic dairy) and in low-grade vegetable cooking oils, purchased by consumers or used in fast food restaurants and school cafeterias. Non-organic cotton is one of the most toxic crops on the planet. Even though 60 percent of what is harvested by weight ends up in the food chain, government regulatory agencies, prompted by large cotton farmers and the garment industry, falsely claim that cotton is not a 'food crop'. This means that super-toxic pesticides and herbicides are allowed to be sprayed in large quantities on the cotton plant. The pesticide residues in cottonseed accumulate in the fatty tissues of dairy cows and are passed on in the milk and dairy products consumed by humans.

Cottonseed oil is routinely laced into a variety of food products, from vitamins to potato chips, and is often added to olive oil without being labelled. This means that GMOs and pesticide residues from cotton crops find their way into a wide range of non-organic food products, triggering health issues including food allergies, cancer, and liver, kidney and immune system damage.

Farmers, farm workers, and residents of rural communities who work and live in closest proximity to cotton fields suffer from exposure to pesticides, GMOs, and chemicals. Many of these agricultural workers suffer from acute pesticide poisoning. Rural cotton farmers, in particular, lack the necessary safety equipment and training for handling the hazardous pesticides, leading to chronic and acute health issues. Pesticides used in cotton farming have been shown to cause endocrine dysfunction, with farmers in rural and poor areas especially at risk.

This cotton is also used in many ways; the least talked about is feminine hygiene products, which are considered 'medical devices'; those who manufacture pads and tampons are not required to disclose their ingredients. Bleached and made from the chemical- and pesticide-drenched materials of non-organic cotton and rayon (wood pulp), pads and tampons contain ingredients that may be toxic and absorbed through skin and mucous membranes.

At a production rate of 25 million tonnes a year, cotton is one of the top four GMO crops in the world. Nearly 95 percent of global cotton production is GMO and/or conventionally grown. Global consumption of non-organic cotton releases large amounts of greenhouse gas into our atmosphere, estimated at about 220 million tonnes a year. One tonne of conventional cotton fibre produces 1.8 tonnes of CO_2e.

The best way to avoid GMO cotton textiles is to buy certified organic cotton and to look for local handspun cotton. Organic cotton is grown without harmful chemicals, leaving the soil, air, and water free from contaminants that cause harm. Organic cotton produces around 46 percent less CO_2e compared to conventional cotton. It also uses far less water to grow since organic cotton growers typically utilize rain far more than irrigation. On top of all that, organic cotton growers use beneficial insects to control unwanted pests instead of relying on harmful chemicals, thus encouraging biodiversity. It is a win-win for humanity and the Earth. It is also known that various skin allergies relate directly to the chemicals used in non-organic cotton farming. Many people with skin problems report a dramatic improvement in their skin condition once they switch to organic. By choosing to buy organic cotton, you are enhancing the health of humans, animals, and natural resources around the world. Supporting organic agriculture is also essential if we want to create improved working conditions for cotton farmers, as farming organic cotton is more regulated and, therefore, fairly traded.

FIBRES OF FREEDOM: THE CHARKHA WAY

In its first year of sales in 2002, Mahyco–Monsanto sold its entire seed stock, with Bt cotton planted on nearly 45,000 hectares. By 2005, on the back of an aggressive marketing campaign, Bt cotton hybrids were planted on more than 500,000 hectares, declaring Bt cotton's success in India. However, the real story unfolding on the ground is very different.

The story of the first four years of Bt cotton farming in India was neatly summarized by P.V. Satheesh, convenor of the Andhra Pradesh Coalition in Defence of Diversity: 'In the first year (2002), Bt cotton was a disaster, yielding 35 percent less than the non-Bt cotton, even while costing four times more than the non-Bt cotton. In the third year, new diseases spread through the soils and the plant. Cattle that grazed Bt cotton plants started dying. And this year [2006], Bt plants have started wilting, forcing farmers to harden their hearts and uproot them. In the village of Mustyalapally, in the Bhongir mandal of Nalgonda, farmers have uprooted Bt cotton from 41 out of the 51 acres planted. The disease has spread to nearby villages, spreading panic among farmers. Farmers complain that the plants are slowly dying one after another because the root system is severely decomposed, without any secondary and tertiary roots on the main root system. Even the bolls formed on these wilted plants did not bear any seeds.'

The story of Bt cotton continued in India, and unfortunately, so did the rise of farmers' indebtedness and suicide. In the cotton belt of Vidarbha, where Bt cotton is widespread, the suicide rate among cotton farmers has spiralled out of control, with reports of more than 100 cotton farmers a month committing suicide in 2006. As a response to the epidemic of farmers' suicides in the cotton belt of Vidarbha, Navdanya launched its 'Fibres of Freedom' campaign. This campaign, which starts with the local cottonseed, is endowed in the Gandhian spirit, and as we follow the 'charkha way', we cannot but help being filled with awe by the foresight Bapu displayed in all the actions that he undertook. Similarly, Navdanya started its work with Vidarbha farmers, sowing seeds of freedom from debt and poison by saving and spreading native cotton varieties of seeds, and creating community seed banks. Following the charkha way, we launched our Fibres of Freedom campaign to link the seed to the loom. To spin those indigenous cotton fibres, Navdanya partnered with the Gandhian organization Magan Sangrahalaya, which adopted the use of natural dyes and natural printing tools, such as leaves, to transform this unadulterated and natural fabric into an equally pure product.

Dyeing Planet vs. Colours of Life

'We don't have to pollute, exploit, and over-consume to look stylish.'

CSILLA HERBSZT

Blogger, The Pretty Planeteer

Colour is the main attraction of any fabric. No matter how excellent its constitution, if unsuitably coloured, it is bound to be a failure as a commercial fabric. In 1856, William Henry Perkin accidentally discovered the colour mauveine while trying to synthesize quinine,[1] and thereby unleashed the era of synthetic dyes and colouring agents; earlier, these were sourced from biodiversity and natural elements.

Every year, mills can use up to 200 tonnes of water per tonne of dyed fabric, which in turn only produces about 1,400 pieces of clothing. Add everything together, and roughly 17–20 percent of industrial water pollution is because of fabric dyes and treatments.

Use of synthetic dyes has an adverse effect on all forms of life. Presence of sulphur, naphthol, vat dyes, nitrates, acetic acid, soaps, enzymes, chromium compounds, and heavy metals like copper, arsenic, lead, cadmium, mercury, nickel, and cobalt, and certain auxiliary chemicals all collectively make the textile effluent highly toxic. Water pollution due to effluents from the textile dyeing industry is a cause of serious concern. The techniques for the detection of dyes are cost-intensive and futile because the dyes undergo chemical changes under environmental conditions, and the transformation products may be more toxic and carcinogenic than the parent molecule. For example, Azo dyes, which account for 60–70 percent of all dyes in the industry, are responsible for setting high-intensity hues, poppy reds in particular. But when broken down and metabolized, they are known carcinogens. And even if it seems like the colour of our clothes and cancer couldn't be less related, Azo and other chemicals don't dissipate but evaporate into the air we breathe or are absorbed through the skin. At best, contact with dyed synthetics triggers allergic reactions, skin irritation, and rashes. At worst, it increases the risk of cancer.

These synthetic dyes produce effluent; if allowed to flow into the fields, it clogs the pores of the soil, resulting in loss of soil productivity. If allowed to flow into drains and rivers, it affects the quality of drinking water in hand pumps, making it unfit for human consumption. About 40 percent of globally used colourants contain organically bound chlorine, a known carcinogen. All the organic materials present in the wastewater from a textile industry are of great concern in water treatment because they react with many disinfectants, especially chlorine. As we have already said, chemicals evaporate into the air we breathe or are absorbed through our skin and show up as allergic reactions and may cause harm to children even before birth.

The tradition of extracting natural dyes from biodiversity is an age-old one, which dates back to 2600 BC, according to the site Quilt History, which mentions a written record found in China to this effect. Reds, browns, and oranges are the first attested colours in a number of ancient textile sites, ranging from the Neolithic to the Bronze Age, across lower Egypt, Mesopotamia, and Europe, followed by evidence of blues and hues of yellows, with green appearing somewhat later.[2] We, therefore, see that the antiquity of natural dyes goes back several centuries, indeed millennia. In fact, it is deemed that vegetable dyes were discovered when hunter-gatherers brushed against berries or flowers and noticed they left their colour on them.

THE MAD HATTER – ALICE IN WONDERLAND

Lewis Carroll's 1865 novel *Alice's Adventures in Wonderland* famously features an eccentric character called the Hatter, who's referred to in the story as 'mad' and became popularly known as the Mad Hatter. However, the phrase 'mad as a hatter' – used to describe someone who is crazy or prone to unpredictable behaviour – didn't originate with Carroll. Instead, the expression is linked to the hat-making industry and mercury poisoning. In the eighteenth and nineteenth centuries, industrial workers used a toxic substance, mercury nitrate, as part of the process of turning the fur of small animals, such as rabbits, into felt for hats. Workplace safety standards often were lax, and prolonged exposure to mercury caused employees to develop a variety of physical and mental ailments, including tremors (dubbed 'hatter's shakes'), speech problems, emotional instability, and hallucinations.

---

1. Quinine is an organic compound that is one of the nucleobases found in nucleic acids.
2. 'Dyes, Colors & Pigments,' Tarek Ismail Kakhia, tarik.kakhia.org

REMEMBERING TOOFAN RAFAI

At the time of the Indian Independence movement, Mahatma Gandhi realized that for the pristine textile of khadi, he needed to adopt natural dyes, to which end he actually wrote a book on it in Gujarati titled *Vanaspatiyon Nu Rang*, originally priced at just six annas. Inspired by meeting Bapu as a young boy from Gujarat, Toofan Rafai became an acclaimed artist and internationally renowned master of vegetable dyes and printing techniques, which he was called upon to popularize through workshops in India and abroad. The 300 shades that he created – many out of waste like onion or pomegranate peel – were, in his own telling, the reflection of a great man's influence on his life: Mahatma Gandhi.

Born in 1921 in the town of Amreli, Gujarat, in a fakir's family, Toofan Rafai got his diploma in drawing and painting from Sir J.J. School of Arts, Bombay, in 1955. A year later, he pursued a postgraduate study in mural decoration, helped along by a scholarship from the Government of India. From 1960 to 1979, Rafai served as the assistant director at Weavers' Service Centre. His work experience comprised several research and development projects. These include innumerable workshops in dyeing and printing using natural dyes at various national and international academic institutions and art centres. Rafai was also a consultant in natural dyes and printing at the National Institute of Design (NID) and the Self-Employed Women's Association (SEWA). In addition, he was a founder member of the Art Council of Indian Artists, New Delhi.

Across the colour spectrum, from red to blue, green, yellow, or black, and shades in between, natural dyes have evolved from animal, mineral, and plant sources. In natural dyes, the easy-to-extract natural colours come from plants, fruits, or flowers.

Traditional Indian Plants for Dyeing

PART OF THE PLANT	PLANT
Roots	Turmeric, madder (manjistha), onions, beetroot.
Bark/Branches	Purple bark, sappan wood, shillicorai, khair, red, sandalwood.
Leaves	Indigo, henna, eucalyptus, tea, cardamom, coral jasmine, lemon grass.
Flowers (Petals)	Marigold, dahlia, tesu, kusum.
Fruits/Seeds	Latkan, pomegranate rind, betel nut, myrobolan (harda).

Natural dyes also have antimicrobial properties, making them safer for kids in particular. Another interesting advantage of natural dyes is that they provide higher UV absorption in the fabrics they are used on. By wearing clothes dyed naturally, you are able to fully protect your skin from the sun's harmful rays. Natural dyes are fungus-resistant and offer excellent protection against antibacterial germs.

Getting Started with Slow Fashion

'Buy less. Choose well. Make it last.'
VIVIENNE WESTWOOD
British fashion designer

While we can all be tagged as consumers, we, as humans, have choices to make regarding *what* we consume and how much we consume of it. In recent years, I have been making efforts to choose conscious consumption wherever possible. If you would like to begin being a conscious or mindful consumer when it comes to clothing, here are some ways to get started:

1. SHOP LESS & BUY LESS
It isn't about buying nothing. Slow clothing slows down the pace of shopping and consumption by asking deep questions that not only slow you down but also inform your choice of consumption. So, buying and shopping less often is your best option to begin this journey. If you must buy something, consider secondhand and vintage. As actor Emma Watson so eloquently put it, 'Every new item of clothing made has a substantial carbon footprint attached to its manufacturing, but the amount of new energy

needed to produce vintage clothing is zero.' Encourage hand-me-down clothing and remove the stigma attached to it. Upcycle clothing or textiles that you already have into something new, like making dresses and skirts out of old sarees, adding embroidery, or a patchwork to jazz up an old shirt or T-shirt. Instead of buying one-time-use clothes for a special occasion, borrow them from a friend.

2. WHEN SHOPPING & BUYING

Remember that your spending is voting for the world you want! So always try and buy handloom, organic, and artisanal clothes, and avoid fast fashion brands. Do your homework to avoid green-washing (brands that say they are slow, ethical, or sustainable in their marketing but don't really live up to it). Look for brand transparency (do they give you all the information about their raw materials, manufacturing, supply chain, pricing, etc.?). Consider smaller, local businesses. Buy higher quality clothes that will last longer, or timeless designs, and seasonless styles.

3. ONCE YOU OWN SOMETHING

If you have less, you can celebrate those few things more. Don't treat anything as disposable. Consider the product's life cycle to keep things out of landfills (repair, donate, upcycle, etc.).

4. DISCARDING CLOTHES RESPONSIBLY

It takes years and decades for most fabrics to break down. Natural materials like cotton, linen, hemp, and silk can biodegrade pretty fast in the right conditions, from a couple weeks to around half a year. Wool and bamboo take longer, often decomposing in 1–5 years. Synthetic fabrics are the kicker – which is what most conventional clothes are made of, such as polyester. It can take anywhere from 20 to 200 years for those to break down! If your clothes are made from natural materials such as cotton, hemp, linen, or silk, you can actually compost them. Just cut the clothing into small pieces after removing buttons, zippers, or non-natural elements. This only works if the clothes are 100 percent natural.

Don't try to compost something that is 70 percent cotton and 30 percent polyester, as the latter is a synthetic fibre. Also, if you use a residential composting service, make sure they accept textiles. Donate, repurpose, and recycle as much as possible.

5. START YOUR STITCHING HOPE COMMUNITY

Build a community to support and promote slow clothing. Gather some friends and start your own Stitching Hope groups. Meet as often as you want and undertake some of these activities together:

- Repurpose your old clothing together.

- Choose quality over quantity – avoid cheap garments made from synthetic fabrics. Educate your group about natural fibres and organize a gathering where you touch and experience these natural fabrics and learn to reconnect to the tactile wisdom of your ancestors. Your family's closets are a great place to start your educational journey.

- Participate in clothes swap events, or exchange old clothing with friends and family.

- Buy natural fibres and learn how to stitch, sew, and repair old garments.

- Learn about traditional weaves and artisans, and local textile heritage.

- Detox your wardrobe, upcycle T-shirts into children's dresses, and old clothes and upholstery into reusable menstrual pads. Find communities to donate all that you are not wearing. Make them beautiful, wash them, iron them, and see someone smile.

- Make eco-friendly detergents and explore natural clothes protection methods, such as dry neem leaves, cloves etc., to store your clothing.

READING LIST

RAHUL JAIN, *Rapture: The Art of Indian Textiles*. Om Books (2011).

ELA R. BHATT, *Anubandh: Building Hundred-Mile Communities*. Sage Publications (2016).

DANA THOMAS, *Fashionopolis*. Penguin Books (2020).

EMMA TARLO, *Clothing Matters: Dress and Identity in India*. University of Chicago Press (1996).

Making Food from Natural Fibre

GONGURA CHUTNEY

HIBISCUS CHUTNEY

INGREDIENTS

Leaves from a big bunch of roselle
1 tbsp sesame, coconut or other oil (plus more for blending)
3 dry red chillies
1 tsp mustard seeds
½ tsp fenugreek seeds
2 cloves garlic
½ inch piece of ginger
1 tbsp of sliced shallot or onion
¾ tsp salt

METHOD

• Wash and dry the roselle leaves. Once the leaves are completely dry, dry-roast them in a flat pan over medium heat in batches. You do not want them to steam but roasted and blackened. Once all the leaves are roasted, sauté them in oil.

• In a pan over medium heat, roast the dry spices (red chilli, mustard seeds, fenugreek seeds) until aromatic. Let them cool a few minutes and then grind them into a fine powder.

• Peel and roughly chop the garlic, ginger, and shallot. Grind all the ingredients including leaves and spices with them in a blender using a little oil. Season to taste with salt. Make sure there is no water used while grinding.

Hibiscus sabdariffa L. is known as gongura in Telugu, lal ambari/patwa in Hindi, and ambadi in Marathi. There are several varieties of hibiscus, some of which are edible and even ideal for making tea. The most commonly used one is Hibiscus sabdariffa L., also known as roselle.

Source: Akshat, *Navdanya*

RICE FLOUR PANKCAKES STEAMED *in* BANANA LEAF
PANKI

INGREDIENTS

1 cup coarsely ground rice flour
¼ cup sour curd
1 tsp green chilli and ginger paste
¼ tsp turmeric powder
¼ tsp cumin powder
½ tsp fruit salt (optional)
2–3 banana leaves
Cooking oil, for greasing

METHOD

- In a medium-sized bowl, combine the rice flour, curd, chilli and ginger paste, turmeric, cumin powder, and fruit salt. Allow it to ferment for 3–4 hours. You can add a little water if the batter is too thick.
- On a frying pan over medium heat, place one greased banana leaf and spread 2 tbsp of the batter over it. Cover with another greased banana leaf and cook on both sides until the banana leaves brown slightly and the pancake in between peels off the leaf easily.
- Repeat with the remaining batter.

India is the largest producer of banana in the world and Asia, and contributes almost a quarter of global production. The banana leaf plate is India's answer to plastics and styrofoam. The fibre is used to make fabrics, now even saris.

PACK A PUNCH *of* HEALTH

FLAX SEED RAITA

INGREDIENTS

1 cup grated bottle gourd
1 cup water
3 tbsp ground flax seeds
1 cup low-fat curd
½ cup finely chopped mint leaves
1½ tsp cumin seeds, roasted and coarsely ground
Salt, to taste

METHOD

• In a deep pan over medium heat, add the grated bottle gourd with water. Cover the pan with a lid and cook for four minutes. Remove from the heat and drain. Set aside.

• In a deep bowl, combine the ground flax seeds, curd, mint leaves, cumin seeds, and the cooked bottle gourd. Season to taste with salt and mix well.

• Refrigerate for at least one hour and serve chilled.

Flax seeds are one of the richest sources of omega 3 fatty acids, essential to build immunity and prevent heart ailments.

7

SLOW KNOWLEDGE

'Real knowledge is to know the extent of one's ignorance.'
CONFUCIUS

Slow knowledge grows through slow living, and in turn, sustains slow living. When we give ourselves time to relate, reflect, experience, observe, and be mindful, our deep knowledge of ourselves, and our awareness of the interconnectedness with other beings grow.

Knowledge comes from the Greek word 'gnosis', a feminine Greek noun which means knowledge or awareness. The root of the word science originates from the Latin 'scientia', which means knowledge. The word intelligence comes from the Latin nouns 'intelligentia' or 'intellēctus', which in turn stem from the verb 'intelligere', to comprehend or perceive, and 'interlegre', meaning, to pick out, or discern. In Sanskrit, 'vijanan' means discernment. Yet, today we may have all the information and data in the world, but discernment and awareness are scarcely practised qualities. We see this in the manner social media is used or fake news is spread by creating a web of misinformation – one that feeds different kinds of violent actions. The ability to acquire knowledge in this new ICE (Information Communication and Entertainment) age is easy and fast, but the ability to discern is only possible when we slow down and contemplate what we are acquiring.

'Fast knowledge', the mechanistic reductionist framework, was knowledge that declared nature as dead matter to facilitate extraction of resources from the Earth and control nature for human exploitation and gain. It had to discount relationships, care, and the value of sacredness, both for animate and inanimate 'things', rather than beings. The founding fathers of this system consciously constructed anthropocentrism based on the assumption of the separation of humans from nature, and the superiority of humans over other species to justify their mastery over nature. Nature was objectified and made an object of manipulation and exploitation. Through knowledge and science, the aim was to transform Terra Madre, Mother Earth, who is living and supports life, into dead inert matter and mere raw material for industrial exploitation.

Many of our current environmental problems arise out of seeing ourselves as separate from nature. Since ancient times, the main goal of science was to gain wisdom and understanding while remaining in harmony with nature. However, since the arrival of Francis Bacon, the goal of science has tended to be patriarchal, and largely involved the pursuit of knowledge in order to control and exploit nature. Dualism, as between mind and matter – championed by the acknowledged father of modern philosophy, René Descartes – sets the paradigm for understanding most of this knowledge system and the culture that it created. Descartes doubted everything until he reached a definite conclusion in his famous dictum '*Cogito ergo sum*' (I think, therefore I am). He deduced that since thought was the essence of nature, mind and matter were separate and distinct entities. Descartes saw the material world as a machine without life or spirit. The natural world functioned in accordance with mechanical laws, and nature could be explained in terms of the mechanistic movement of the parts. Even human beings belonged to a category of machines, in which the human body was seen as a container activated by a soul that was connected to the body via the pineal gland in the brain. Thanks to Cartesian dualism, the mechanical view of nature became the dominant view of classical science. Isaac Newton used differential calculus to come up with a mathematical formulation that undertook Descartes' work and completed the mechanical worldview. For Newton, God had set the whole universe in motion, and it has continued to run ever since like a machine governed by immutable laws. The view is essentially deterministic and fatalistic. It says that given sufficient knowledge of nature's laws, we could have predicted the present. The future is already cast, and free will is an illusion.

On the other hand, ancient, intricate systems of traditional indigenous knowledge, which are part of the slow knowledge system, are based in the celebration of biodiversity. The knowledge has traditionally been freely available within and between communities in the Commons (lands, water, air, and forests). In this system, biodiversity has intrinsic value as well as high use value. For commercial interests, on the other hand, biodiversity itself has no value; it is merely 'raw material' for the production of commodities, for the 'mining of genes and intellectual property', and for the maximization of profits.

Fast vs. Slow Knowledge

'What we observe is not nature itself, but nature exposed to our method of questioning.'
WERNER HEISENBERG
Physics and Philosophy: The Revolution in Modern Science

Critiquing this mechanistic worldview, physicist Fritjof Capra argues that this classical notion is now receding. According to him this model has a number of entrenched ideas and values that have recently been brought into question, such as the view of the universe as a mechanical system composed of elementary building blocks, challenged by Einstein's theory of relativity; the view of the human body as a machine having different parts to be fixed separately; the view of life in society as a competitive struggle for existence challenged by the web of life; the belief in unlimited material progress to be achieved through economic and technological growth challenged by the current climate crisis – all of this linked to the belief that a society in which the female is subsumed under the male follows a basic law of nature that modern physics contradicts today.

So, let us look at the key differences between Fast and Slow Knowledge.

FAST KNOWLEDGE	SLOW KNOWLEDGE
Anthropocentric, or human-centred, where human beings are masters of nature.	Grounded in an ecocentric worldview. Humans are not separate from nature. The world is not a collection of isolated objects, but is a network of phenomena that are fundamentally interconnected and interdependent.
Survival of the fittest.	Interconnection and adaptability, with empathy and collaboration at the core of survival.
Thinking is rational, based on analysis, reductionist, and linear.	Thinking is intuitive, synthesis-oriented, holistic, and nonlinear.
Scientific facts are independent of what we do and are therefore independent of our values.	Values are not peripheral to science, nor to its applications to technology, but constitute their very basis and driving force. Scientists, therefore, are responsible for the research, not only intellectually but also morally.
Values driving fast knowledge are expansion, competition, quantity, and domination.	Values reflected in slow knowledge are conservation, cooperation, quality, and partnership.

Hierarchical knowledge systems are driven by specializations, with Western knowledge systems given pre-importance. Knowledge and therefore power emerges in such a system as top heavy and concentrated in the hands of a few.	Knowledge is decentralized and viewed as a network of relationships and patterns, creating a social network; people are empowered by being connected to the network. Power as empowerment means facilitating this connectedness. Indigenous and local wisdom is not devalued.
Reactive and dualistic, leading to polarization and subjugation.	Reflective and holistic, leading to collaboration and healing.
Mega-data driven, where data is absolute and restrictive, to be mined for profit.	Information and discernment as symbiotic, and data for the sake of data without reflection of the impact on interbeing is questioned.
Banking system of education.	Conscientization by Paulo Freire. Taken from the Portuguese term 'conscientização', which allows one to learn social, political, and economic incongruities as a necessary step to act against oppressive elements that cause inequalities.
Education monocultures and elitist universities.	Diversity and Earth University.

Protection of Indigenous Rights and Biodiversity

'We are not myths of the past, ruins in the jungle, or zoos. We are people and we want to be respected, not to be victims of intolerance and racism.'
RIGOBERTA MENCHÚ TUM, K'ICHE'
Indigenous feminist & human rights activist, Guatemala

There are two paradigms of biodiversity conservation. The first is held by communities whose survival and sustenance are linked to local biodiversity utilization and conservation. The second is held by global commercial interests whose profits are linked to utilization of global biodiversity for the production of inputs into large-scale homogenous, uniform, centralized, and global production systems.

Even today, most people in India live in the first paradigm of biodiversity utilization. According to an ethno-botanical survey, there are 7,500 species used as medicinal plants by the indigenous medical traditions of India. These traditions are kept alive by 360,740 Ayurveda practitioners, 29,701 Unani experts, and 11,644 specialists of Siddha. In addition, millions of housewives, birth attendants, and herbal healers carry on village-based traditions.

In the 1990s, before the impact of globalization, 70 percent of healthcare needs in India were still based on traditional beliefs and systems based on the use of medicinal plants. As 80 percent of seeds used by farmers still come from farmers' seed supply, India is therefore still a predominantly biodiversity-based knowledge economy. The impacts of globalization have eroded the biodiversity and knowledge sovereignty of local communities, pushing them into deep poverty and unemployment.

In 1992, the international community adopted the Convention on Biodiversity (CBD) at Rio de Janeiro at the Earth Summit, along with the United Nations Framework Convention on Climate Change. These two environmental agreements are the fragile thread holding together our future and the diversity with which we perceive and interact with the world.

The CBD introduced during the 1992 Earth Summit was seen as a way to reclaim the lost ground of unchecked exploitation of genetic resources in the developing countries. While no mandatory obligations arise out of this legal instrument, it lists three important aims:

1. Conserving biological diversity
2. Sustainable use of resources
3. Fair and equitable sharing of benefits that arise out of commercial use

Though it is still a form of Northern (First World) 'takings' with which a 'giving' to the South (Third World) is attached, in aiming for justice and fairness, the CBD mandates sharing of benefits with resource providers. CBD has carved an institutional framework for benefit-sharing arrangements which have sprouted across the world, through which the resource users (corporations and scientists) compensate the resource providers and stewards (indigenous people, communities, etc.) through the idiom of 'benefits'.

The CBD reaffirmed the recognition of *'the intrinsic value of biological diversity and the ecological, genetic, social, economic, scientific, educational, cultural, recreational, and aesthetic values of biological diversity and its components.'*[1]

The primary objective of the Convention is the conservation of biodiversity. While recognizing that conserving biodiversity is a common concern for humanity, the Convention clearly recognized the sovereign rights of states and closed the door for unregulated access for exploitation and ownership through patents by commercial interests and global corporations. It strengthened the sovereign option in the area of biodiversity and closed the colonial option of 'free access' to our biodiversity to be sold back to us as patented commodities for health and agriculture.

The Convention also recognized that indigenous knowledge needs to be protected in order to conserve biodiversity. Article 8(j) on *in situ* conservation states:

> Subject to its national legislation, respect, preserve and maintain knowledge, innovations and practices of indigenous and local communities embodying traditional lifestyles relevant for the conservation and sustainable use of biological diversity and promote their wider application with the approval and involvement of the holders of such knowledge, innovations and practices and encourage the equitable sharing of the benefits arising from the utilization of such knowledge, innovations and practices.

On 29 October 2010, the Nagoya Protocol was adopted in Nagoya, Japan, with the aim of establishing an international and legally binding framework for transparent and effective implementation of the Access and Benefit Sharing (ABS) concept at the regional, national, and local levels.[2] It was a movement forward in cementing a new vocabulary of biocultural entitlements, where it called for the creation of set duties and obligations on the parties engaging with indigenous communities for the use of genetic resources and knowledge.

It noted in its Preamble, '*the interrelationship between genetic resources and traditional knowledge, their inseparable nature for indigenous and local communities, the importance of the traditional knowledge for the conservation of biological diversity and the sustainable use of its components, and for the sustainable livelihoods of these communities.'*

It further stated that it was mindful of '*the right of indigenous and local communities to identify the rightful holders of their traditional knowledge associated with genetic resources, within their communities.'*

In the domain of benefit sharing, the Nagoya Protocol thus institutionalized various biocultural rights and entitlements of indigenous communities in accordance with the domestic laws of the parties. However, this Protocol restricts access only to global players, ignoring the access of local communities. It prioritized only utilization for research and commerce and ignored the survival needs of local communities. This was legalized biopiracy, as it enabled the transfer of genetic wealth from local communities to global corporations. It undermined the biodiversity economies and cultures that have conserved biodiversity and are necessary for conserving it for the future.

However, despite that, the Protocol successfully provided a legal framework for the fair and equitable sharing of benefits arising out of the utilization of genetic resources for research and commercialization purpose. At present, more than a 100 countries have acceded to the Protocol. This makes it mandatory for them to set down an access and benefit regime in compliance with the Protocol. Furthermore, patent and IPR regimes need to respect the obligations under the Nagoya Protocol for access and benefit sharing.

While we worked on laws to protect our biodiversity, we also built the movement to defend the biodiversity and knowledge commons within communities.

---

1. Preamble of Convention on Biological Diversity; available at https://www.cbd.int/doc/legal/cbd-en.pdf.
2. Nagoya Protocol; available at https://www.cbd.int/abs/doc/protocol/nagoya-protocol-en.pdf.

Liberating the 'Free Tree': Challenging the Neem Biopiracy Patent

To forget one's ancestors is to be a brook without a source, a tree without a root.
CHINESE PROVERB

In 1971, US timber importer Robert Larson observed the neem tree's usefulness in India and began importing their seeds to his company headquarters in Wisconsin. Over the next decade, he conducted safety and performance tests on a pesticidal neem extract called Margosan-O, and in 1985, he received clearance for the product from the US Environmental Protection Agency (EPA). Three years later, he sold the patent to the multinational chemical corporation W.R. Grace and Company.

The botanical name of the neem tree is *Azadirachta indica*, which is taken from the Persian name for the tree, Azad-Darakth, meaning 'the free tree'. India has at will shared its 'free tree' and knowledge of its myriad uses with the world, which is why it now flourishes in many countries of Africa, Central and South America, the Caribbean, and Asia. Neem has been used for diverse purposes over centuries, primarily in medicine and agriculture. Research has shown that neem extracts can influence nearly 200 species of insects, many of which are resistant to pesticides. The medicinal properties of neem have also been described in ancient Indian medicinal literature (*Atharva Veda*, the *Ghthyasutras*, and the *Sutragranthas*). In fact, the Sanskrit name *nimba* is a derivative of the term *Nimbatisyasthyamdadati*, which means 'to give good health'. Unani texts detail the pharmacological action and the therapeutic efficiency of neem, and Unani scholars consider that neem can take care of nearly all aspects of human health, and recommend its widespread use, as it is available free of cost.

However, the product gifted by neem almost became the 'intellectual property' of Western scientists and large corporations by being patented. As early as the 1970s, scientists in the USA and in Japan filed patents in their countries on the various properties of the neem and their extraction processes, which is ironic since these principles and properties have been known and used for centuries in India. Even the processes that were claimed to be patented were only minor modifications of those that have been used for centuries to prepare extracts.

Since 1985, multiple patents have been filed by US and Japanese firms on the formulae for stable neem-based solutions, emulsions, and toothpaste. At least four of these patents were owned by W.R. Grace, three by another US company, the Native Plant Institute, and two by the Japanese Terumo Corporation, before they were revoked in the court of law.

In response to the patents in 1995, the Research Foundation for Science, Technology and Natural Resource Policy, in conjunction with the International Federation of Organic Agriculture Movements (IFOAM) and 200 other associates, legally challenged two of the neem patents held by W.R. Grace, one in the US and the other in the European Patent Office (EPO).

The sole objective was to highlight the issue of 'biopiracy' and emphasize the need to create a legal system that takes full cognizance of the innovation in indigenous systems of knowledge and treats the developments built on this knowledge as obvious steps, given that the knowledge of the utilization of biodiversity already exists.

As part of the Neem Campaign, a massive signature campaign was also launched in India demanding that the patents on neem be withdrawn. More than 100,000 people – children, students, housewives, professional workers, doctors, lawyers, teachers, farmers, engineers, environmentalists, and artists – joined the movement to protest these patents. The signatures collected were finally submitted on 9 May 2000 in Munich, Germany, a day before the revocation of the patent, to the EPO by Dr Vandana Shiva and Afsar H. Jafri of Research Foundation for Science, Technology and Ecology, along with Mr Nammalwar, an Indian farmer, and the Indian scientists Prof Udai Pratap Singh and Dr Abhay D. Phadke. History was made on 10 May 2000.

Mr Tzschoppe D., the chairman of the Opposition Division of EPO, declared that the 'patent is revoked and the proceeding stands terminated'. The four-judge panel revoked in its entirety the patent number 436257, which had been granted to the USA and the multinational corporation W.R. Grace for a fungicide derived from neem seeds.

Since then, Navdanya has successfully fought and won patents on Basmati and turmeric.

5 WAYS TO SLOW KNOWLEDGE

1. MAKE YOUR PEOPLE'S COMMUNITY BIODIVERSITY REGISTER

A People's Community Biodiversity Register (PBR) is the documentation of the resources and knowledge of local communities at the local, regional, and national levels by the people themselves. This is done for the purpose of rejuvenating the ecological basis of agriculture and the economic status of the farmers.

The documentation of community knowledge is imperative for the following reasons:

i. EROSION OF RESOURCES

Non-sustainable production and consumption patterns in agriculture have led to the erosion of land, water, and agricultural biodiversity in farmers' fields. For example, the 'miracle seeds' of the Green Revolution replaced indigenous varieties of rice, and many other crop seeds like the amaranth, which are in the process of being replaced by crops like rice and wheat; they are thus threatened by extinction.

ii. EROSION OF KNOWLEDGE

Communities that are identified and innovative have traditionally had free exchange of knowledge of their resources within the community and outside it. When such resources are eroded and lack common use, common knowledge gets eroded over time.

iii. DISAPPEARANCE OF SUSTAINABLE UTILIZATION OF ALTERNATIVES

When both the resource and knowledge about it disappear from the Commons, the space for utilization of alternatives in a sustainable manner shrinks, as the space for a return to sustainable agricultural production and consumption lessens.

iv. INTELLECTUAL PIRACY

The removal of knowledge from the Commons leaves it vulnerable to being claimed as someone's private intellectual property. This is particularly true when the common knowledge has no recorded originator or innovator but has traditionally been treated as community knowledge. The intellectual property right (IPR) regimes thus ensure that the pirates of such knowledge become the new owners of the knowledge and share it only for profit.

v. BIOPIRACY

Intimately linked with intellectual piracy is biopiracy. The removal of resources from the Commons leaves it vulnerable to piracy, both by the IPR regimes and by collections made by organizations (nationally and internationally, government or private).

vi. IPRs AND MONOPOLIES

Together, intellectual piracy and biopiracy mean that the resource is now in the monopoly control of corporations. In agriculture, this reduces all innovation to innovation by the corporations for profits, additionally resulting in agricultural production and consumption becoming conditional on corporate interests.

BIODIVERSITY REGISTER

In an attempt to preserve India's vast biodiversity, the government enacted the Biological Diversity Act, 2002. Besides provisioning for a three-tier administrative structure in the form of the National and State Biodiversity Boards, along with Biodiversity Management Committees (BMC) at the local level, the Act also adopted the concept of the PBR to record local and traditional knowledge on biological heritage.

2. SET UP GRANDMOTHERS' UNIVERSITY

The wisdom of grandmothers is in their capacity to love unconditionally; stemming from this love is our food culture, healing practices, and our ability to discern, which have been developed through dialectical storytelling. In the current worldview based on competition, insecurity, and fear, the steadiness of love and compassion is brought to the next generation through this wisdom of our grandmothers. While it is called sustainability in today's jargon, it is really thinking about future generations – not just of me, myself, today. We need to create grandmothers' universities everywhere so that this local knowledge never disappears. Grandmothers and the elders are the best link and the true source of biodiversity and historical wisdom of human culture.

Michael Pollan, US journalist and author has said, 'Don't eat anything your grandmother wouldn't recognize.' What we eat offers both sustenance and an awareness of how the Earth supports us individually and collectively. When food is sacred, it is medicine for our bodies, minds, and spirits. We support communities in keeping alive their traditional sources of nutrition and healing.

The Grandmothers' University – or intergenerational gathering with an aim to both celebrate and validate the wisdom of our grandmothers, as well as transmit this to future generations to arrest the rapid erosion of skills, knowledge, and values – evolved by women over millennia to live sustainably, is a survival imperative as we face the vagaries of climate change.

3. MAKE YOUR OWN WISDOM RECIPE BOOK

You can divide your recipe book into four broad categories:

- Healing Food – Home Remedies (*Nani ke nuskey*)
- Festival Food
- Daily Diet
- Sacred Food

4. LEARNING TO DISCERN: SLOWING DOWN THE INFORMATION HIGHWAY IN THE ICE AGE

The world is riding a bullet train on the information highway. Information Communication Entertainment (ICE) governs every aspect of our daily lives – from the moment we wake up to the time we go to bed. There is an information overload that overwhelms us, sometimes paralysing us into apathy, and at times catapulting our emotions into knee-jerk reactions. Thus, it becomes essential that we learn the art of discernment by practising stillness and reflection. As Gandhi said, action without reflection is merely re-action. So, what can we do to navigate this tsunami of information?

To encourage people to think more critically about the media they are consuming, the **Centre for Media Literacy** developed these five questions one should consider:

i. Who created this message?

ii. What techniques are used to attract my attention?

iii. How might people understand this message differently?

iv. What lifestyles, values, and points of view are represented in, or omitted from, this message?

v. Why was this message sent?

By asking these questions regularly about the messages you see, read, and hear, you can become more media literate, better formulate your own opinions, and better express these opinions in public debates and discussions. You can visit their website for more tools and a better understanding of how to navigate this labyrinth of information that one finds oneself often lost in.

5. GYAN SWARAJ: CHALLENGING LARGE DATA MINING

The 'death of nature' assumption is at the core of the logic of extractivism and the metaphor of mining – of land from indigenous peoples, of fertility from the soil, of water from rivers and underground aquifers, of genes from biodiversity, and knowledge from indigenous communities. Biopiracy is the mining of knowledge and biodiversity for patents and intellectual property rights.

A new biopiracy is underway – through patents on data from our bodies and minds, and mining of such data as 'Human Body Activity'. We are being turned into the new raw material. Our minds and bodies are the latest colony for mining and extraction. They did say 'data is the new oil', and just as the oil industry extracted oil to fuel its war on the planet, data is already being used against the minds and bodies of people. We must stay alert when we are pulled more and more into the surveillance-obsessed society. Here are some simple steps one can take to challenge data mining:

i. Check your digital footprint periodically and ensure that your personal data is not being used to target you with propaganda and a consumerist lifestyle.

ii. Question 'scientific practices' and creations, especially medicine- and food-related discoveries, to see if they are in fact safe for both you as well as the planet. Do adequate research and discussion when ingesting or taking anything that is being peddled for health reasons. Medicines and food supplements that have been fortified or genetically modified deserve more scrutiny.

iii. Surveillance society needs to be questioned, as it links to your democratic freedom and also makes you vulnerable to state and transnational control, so always be vigilant and don't give up your personal information that easily.

THE NAVDANYA WAY

The Grandmothers' University was started by Navdanya in 2007 both as a programme of Bija Vidyapeeth (Earth University, Dehradun) as well as a grassroots movement documenting traditional knowledge and the use of biodiversity through Community Biodiversity Registers – a process which began in Pattuvam, a village in Kerala in 1997.

Many of our best childhood memories are connected to eating with our grandparents, and we need to ensure that we do not deprive our children of this joy. If children grow and cook food, they will eat it. To facilitate food knowledge and legacy, Navdanya started Little Chef programmes. Aimed at schools, the programme was to teach children to grow a Garden of Hope. Alongside, there were other programmes for learning to cook and connect with food history. Sharing the template of collecting recipes, the programme aims to document the wisdom of family elders and help keep the kitchen a space of living knowledge.

Little Chef Programme Recipe Template
(print and use)

Name of the Recipe:

Prep time:

Person who gave the recipe:

Ingredients:

Photograph

Method:

Story:

Benefits:

Season to enjoy the yumminess:

GARDEN OF HOPE: BECOME A SEED KEEPER

We learn from the seed, its generosity, and its diversity. You have just celebrated the harvest in your kitchen garden. Now it's time to take your learning and savings for the next harvest cycle. The knowledge of the garden is all stored in nature's microchip, the seed. Let's learn to save and protect them for the next harvest cycle.

GUARANTEE GENETIC PURITY WITH...	CONSERVE PHYSICAL PURITY WITH...
Taking care of the source of the seeds.	Taking care while harvesting the crops, collecting the seeds and threshing.
Maintaining isolation distance.	Rotating your crops.
Avoiding cross-pollination by removing off-types before they reach the flowering stage.	Drying and storing the seeds well.

1. COLLECT THE SEEDS

The collection of the seeds differs from one plant to another.

2. DRY THE SEEDS

The grains should be dried in the sunlight. For drying, lay the seeds in the sun on a mat or a plastic sheet, and spread them thinly. Mix and turn the seeds 4–5 times a day.

3. CLEAN THE SEEDS

Once dried, clean the seeds by removing all the stones and the broken, malformed, undersized, or diseased ones.

4. STORE THE SEEDS

Challenges of seed storage are to regulate moisture and avoid pests. Traditionally, people used airtight containers made of bottle gourd, wood or bamboo if possible, covering them with cow dung and mud to avoid pests. According to the resilience of the crops, anything, from glass to bamboo containers and gunnysacks, can be used to store seeds and control moisture.

5. CONTROL THE PESTS

To control the pests, you can add the following components to the container:

- **Neem**: Neem leaves are dried in the shade and then powdered. This powder is added to the bags or containers in which the seeds are kept. Using neem oil is easier and more efficient. Do note that the quantity of oil used should be one percent of the weight of the seed.

- **Garlic powder**: Add 1 or 2 garlic cloves in 2 kilograms of seeds. Or, crush one clove of garlic and put it with the seeds. In some regions, an entire onion is also put along with the seeds.

- **Wood ashes**: Add 25 percent of ash to the container. For every kilogram of seed to be stored, add 500 gms of fresh dry-wood ash that has cooled. After the seeds have been put in the container, add more ash to cover the seeds.

- **Lime**: For every one kilogram of seeds stored, add 50 gms of lime.

You can also plaster the room with cow dung where seeds are stored to keep away pests. Finally, take care to write on the container the name of the plant, the place of the collection, and the time of the collection.

READING LIST

VANDANA SHIVA AND KARTIKEY SHIVA, *One Earth, One Humanity vs. the 1%*. PM Press (2019).

SATISH KUMAR AND LORNA HOWARTH, *Regenerative Learning: Nurturing People and Caring for the Planet*. Schumacher College (2022).

MO GAWDAT, *Scary Smart: The Future of Artificial Intelligence and How You Can Save Our World*. Pan McMillan (2021).

Recipes to Celebrate Indigenous Knowledge

MODUR PULAV (SWEETENED RICE)

INGREDIENTS

5 cups milk
300 gm Basmati rice, washed and soaked for 1 hour
2 cups sugar
2 gm saffron
5 tbsp ghee
½ cup almonds
¼ cup cashew nuts
¼ cup raisins
¼ cup grated dry coconut
3 cinnamon sticks (1 inch each)
6 green cardamoms
6 black peppercorns
2 bay leaves
6 cloves

METHOD

- Heat the milk in a heavy-bottomed pot over medium heat. Add the drained rice and cook over low heat until the milk begins to get absorbed. Remove from the heat, keeping the rice a little underdone. Transfer the rice to a large flat dish. Add the sugar and mix well gently.
- Soak the saffron in ¼ cup hot water for 5 minutes.
- Heat the ghee in a heavy-bottomed pot over medium heat. Lightly fry the dry fruits and coconut, cinnamon, green cardamoms, peppercorns, and bay leaves.
- Add the rice and saffron water. Stir well to combine. Place a flat pan/griddle under the pot. Cover the pot with a lid and cook over low heat until done. Garnish with cloves and serve hot.

HALDI DOODH
(AKA TURMERIC LATTE)

INGREDIENTS

1 cup milk (organic preferably)
1 tsp turmeric powder or 1 inch-long root of turmeric
A pinch of crushed pepper, black or white (alternatively, use a few whole peppercorns)
1 green cardamom, crushed
1 tsp (or more as per taste) of powdered gur /jaggery (organic preferably); those who prefer sugar can use brown sugar

METHOD

- In a saucepan over low-to-medium heat, add the milk and bring to a rolling boil.
- Add the turmeric, pepper, and green cardamom; let the mixture simmer, stirring it off and on for a few minutes before removing from the heat.
- Add the jaggery or brown sugar; if you want to use honey, allow the milk to cool down a bit before adding.

SUMMER VERSION

- Along with the same ingredients, also use 3–4 blanched almonds, skin removed and ground to a paste.
- The mixture should then be cooled to room temperature and refrigerated to become cold, but not chilled. Some people add 1 teaspoon of ghee or coconut oil to this latte, as the addition of ghee is soothing to the throat.

This preparation is deemed anti-ageing and anti-allergic. Milk is soothing; turmeric is anti-inflammatory and combats plaque in the brain that could lead to Alzheimer's.

Winter Version
adapted from www.mapi.com

HEALING
METHI MUTHIA

INGREDIENTS

1 bunch fenugreek leaves
8 green chillies
1-inch piece ginger
½ tsp cumin seeds
1 cup sorghum flour
¼ cup chickpea flour
¾ cup whole-wheat flour
½ tsp coriander powder
Salt, to taste
½ cup cooking oil
Juice of 1 lemon
1 tsp sugar
Water, for kneading

METHOD

- Wash the fenugreek leaves well and shake off the excess water. Finely chop and keep aside.
- In a mixer-grinder or spice grinder, grind the green chillies, ginger, and cumin seeds. Transfer to a large plate with a rim.
- Sift the chickpea flour, sorghum flour, and wheat. Stir into the green chilli/ginger mix. Now add the coriander powder, salt, oil, lemon juice, and sugar. Add water to make a stiff and smooth dough. Form into lemon-sized balls, flatten a little with the hand and steam for 10 to 12 minutes.
- Serve warm with a spicy or sweet chutney of your choice.

8

SLOW ECONOMY

'It seems madness to think that a society would rate marginal economic growth above a livable earth, but there you are. I had always assumed the reason to build a bigger economy was to make the world a better place. In fact, it appears, the reason to build a bigger economy is, well, to build a bigger economy.'

BILL BRYSON

American journalist and author

Slow economies are living economies. As the art of living, the economy is aligned with nature and society's life-giving processes. It is based on the worldview that the Earth is for all beings, on the recognition of and respect for the ecological limits of nature, and on the rights of all human beings. Living economies sustain life, as the capital here is relationships and the currency is care. This cultivates the art of living, not of running fast and non-stop in a money machine.

The word resource is derived from 'resurge', meaning 'that which renews'. With industrialization, this word has been reduced to 'raw material'. In slow economies, the Earth is living; her resources are living and renewable. In the extractive economy, on the other hand, nature is dead, inert matter, and resources get 'value' only when exploited as raw material for profit through industrial processes. In such economies, nature becomes poorer as ecological crises increase. Poverty in the world, seen as growing hunger and disease, homelessness, displacement, wars and refugees, is thus a result of not giving back to nature and society, connected through the web of life.

Value comes from the Latin word 'valere', meaning 'be strong, be well; be of value, be worthy'. In slow economies, value is lived and experienced. It rises from within and isn't assigned externally. All beings belong to the Earth. They all have value, and their value is their well-being.

The original meaning of wealth was not money but 'well-being'. In slow economies, we reclaim wealth as well-being by engaging in the right action and doing the right thing for all our human and non-human connections.

Fossil fuels speed up output but degrade the planet and displace people. This speed, in the name of efficiency and productivity, ruptures nature's cycles and limits and violates planetary boundaries, creating waste and pollution and emitting toxins in the environment, our body, and mind. All ecological crises are the rupture of nature's cycles, and the transgression of what has been called planetary boundaries.

Nature knows no waste or pollution. Slow economies mimic nature and are zero-waste economies. These are fossil-fuel free, chemical-free, and plastic-free. They celebrate the handmade and artisanal, helping us shift from being consumers to creators.

Living and circular economies are aware of and maintain nature's cycles by giving back to the Earth. They work in accordance with nature's ecological cycles and rhythms; based on giving, on what is referred to as the 'Law of Return'. When we give back to maintain the living systems in nature and society, we create circular economies. The Earth gives us food. When we give back part of her organic gifts to the soil, she continues to give us food and not poison. Wealth is shared, and as it circulates, it does not concentrate in a few hands. In these economies, the parameter of growth is well-being, of all animate and inanimate beings, and a gift of wellness for all humanity, as it celebrates the interconnection and rhythm of the living Gaia. In the care of the Earth and society, diversity of meaningful and creative work is possible. It is based on nature's Law of Return. When economies are circular, every living being and every place is the centre of the economy, and nature and society evolve and emerge from multiple self-organized systems, like the trillions of cells in our body.

Slow economies are based on biological and cultural diversity, and they nourish, conserve, and regenerate diversity in nature and in society. Circular economies, as living economies, are by their very nature biodiverse, spanning from the intimate and local to the global and planetary.

Slow Economy vs. Fast Economy

'Isolation and loneliness are central causes of depression and despair. Yet they are the outcome of life in a culture where things matter more than people. Materialism creates a world of narcissism in which the focus of life is solely on acquisition and consumption. A culture of narcissism is not a place where love can flourish.'
BELL HOOKS
All About Love: New Visions

The economy, which is part of society, has been placed outside and above society, beyond democratic control. Ethical values, cultural values, spiritual values, and values of care and cooperation have all been sidelined by the extractive logic of the global market that seeks only profit. Competition leaves no room for cooperation. When reality is replaced by abstract constructions created by the dominant powers in society, manipulation of nature and society for profits and power becomes easy. The welfare of real people and real societies is replaced with the welfare of corporations. The real production of the economies of nature and society is replaced by the abstract construction of capital. The real, the concrete, the life-giving gives way to artificially constructed currencies, fertilizers, food, and even artificial intelligence, thus concentrating ownership and power in the hands of a few.

The dominant model of the economy no longer has its roots in ecology but exists outside and above ecology, disrupting the ecological systems and processes that support life. The unchecked conquest of resources is pushing species to extinction and has led ecosystems to collapse, while causing irreversible climate disasters.

The money machine is a linear, extractive, violent machine. It extracts resources, minerals, soil fertility, and water from the Earth. It mines and pirates genes from seeds and life forms. It is now 'mining data' from our farms, relationships, and lives. It transforms every Commons into a commodity to be sold for profit. By constantly enclosing the Commons that sustain life and create real growth, it creates 'growth' measured as money-making, not growth in life, well-being, and happiness. This system, based on linear growth, causes disruption of natural cycles and creates the crisis of hunger, health, and ecological collapse, where happiness and human purpose are reduced to money and how fast one can earn it. The non-reflective and fast-paced need to consume everything, along with the human identity being reduced to that of a mere consumer, robs us of our human experience, commodifying life itself.

FAST ECONOMY	SLOW ECONOMY
Economy of greed, 1 percent (wealth in the hands of few).	Economy of care, 99 percent (money works for everyone).
Fast money.	Slow money.
Anthropocentric.	Earth-centric.
Economy of desires.	Economy of simplicity.
Reactive markets.	Responsive markets.
Privatization of the Commons.	Custodians of the Commons.
Monopolies of ownership.	Sarvodaya (You are, therefore I am, and we rise together).
Technological feudalism (data mining and ownership of discourse, habits, and political power in the hands of big tech).	Technology of care (based on bio-mimicry and respecting the web of life).
Economic growth linked to concentration of wealth and 'money'.	Growth as well-being for all, anchored by equality, compassion, and mindful living.

Reimagining Growth, the Circular Economy Way

'No no no
Don't it always seem to go,
that you don't know what you've got
Till it's gone
They paved paradise
And put up a parking lot
Hey farmer, farmer
Put away the DDT
I don't care about spots on my apples
Leave me the birds and the bees
Please!'
'BIG YELLOW TAXI'
Counting Crows

Growth is a biological term. A seed becoming a tree is growth, with the tree staying rooted to where the seed was planted. But in the representations of growth in the financial world today, you see illustrations of plants growing out of coins, and their leaves drawn like coins or currency notes. This is a misplaced metaphor, suggesting money grows on trees and anyone who works hard, like the farmer on land, can grow a pot of coins for themselves.

But this is far from the truth, as currently 'the richest 1 percent grabbed nearly two-thirds of all new wealth worth $42 trillion created since 2020, almost twice as much money as the bottom 99 percent of the world's population,' reveals a new Oxfam report, 'Survival of the Richest: The India Story' (2023).

The contemporary global economy is based on the reinvention of the colonizing project. That is exactly how it works. It is the colonizers who define the historic narrative, writing the laws and the rules which serve to legitimize the sacking of lands, resources, and wealth perpetrated against the colonized. The first colonization established the concept of Terra Nullius – 'empty land' – to seize the territories of colonized populations and make them into the property of the colonizers. In the contemporary world, the biotechnology and chemical industries have constructed the concept of Bio Nullius – 'empty life' – to steal seeds and try their hand at biopiracy with the use of patents and intellectual property rights. The tech giants and surveillance capitalists, such as Google, Facebook, and Microsoft, have established the concept of Mens Nullius – 'empty mind' – to control and take possession of our minds and of our lives.

The period after the Second World War was hailed as the time of decolonization and liberation of many countries. However, two things took place during this period that just reinvented colonization. One was that Hitler mobilized companies like Bayer, BSF, and IG Farben to make chemicals to kill people, which we use today as synthetic chemical fertilizers (and fertilizer bombs). Similarly, pesticides were introduced that are basically gases like Zyklon B that were used in concentration camps to gas people. The second part of the war was transforming the concept of growth into that of making money, because governments had to mobilize more money to finance wars. Even today, according to the United Nations Systems of National Accounts, if you consume what you produce, it is not counted as production. Only if you sell or buy a product is it counted. This system has killed the circular economies of sustenance, such as that of nature, women's work, and local communities.

The second problem emerged with the deregulation of most financial systems across the world in the late 1980s and early '90s. So far, growth had been created by extracting money from real economies. Now, money could make money in fictitious economies. It has no connection with the real world. The top billionaires today don't produce anything real. They control platforms, your knowledge, and also your mind – whether it is Facebook, Amazon, Microsoft, or X (formerly Twitter). This system has worked a way to make the value of what farmers grow reduce to zero. But the idea of limitless growth in a world of ecological limits is an impossibility and also reproduces a malfunctioning biological system. The only cell in our body – if we have it – that doesn't know when to stop growing is cancer, and it kills. So the economy, having been reduced to fictitious growth and extractive growth, is a cancer that is killing the planet, the society, and livelihoods. The two big crises of our times are the ecological crisis and planetary collapse, and the poverty crisis of large numbers having no means of livelihood or resources. And it will only get worse if this artificial construct is what rules the world.

All living systems have limits. The cells in our body are decaying and replacing themselves every day. Trees in a forest die, some new ones are born, and the forest keeps replenishing itself to compensate for the lost. It is the cycle of birth, maintenance, and collapse. The trajectory of linear, limitless growth fails to see this. So, we need to give primacy to living systems – living societies, living cultures, living minds, living ecosystems – and stop giving primacy to fictions that kill.

To embrace the circular economy, there is a need to understand the key myths propagated by the linear economy and how it impacts our understanding of living in this world. Here are five myths about the economy that we must consider as we build circular living economies of care:

MYTH 1: THE ECONOMY IS SEPARATE FROM NATURE

The word economy has its roots in the Greek word 'oikos', which originally referred to the family household, its daily operations and maintenance. In 1988, Ernest Herschel, the leading German disciple of Charles Darwin, derived the new label 'oecologies' (from the same root word 'oikos') to refer to the science of the relations of living organisms to the external world, that is, their habitat, customs, and energies.

The economy is dependent on nature, where nature is not subordinate to 'human' needs but humans are part of the ecological system. The historic dependence on fossil fuel-driven economies, which has led to numerous wars and ecological collapse, manifested in the climate crisis, shows us what this 'separateness' from nature is hurtling us towards.

MYTH 2: THE ECONOMY IS ABOUT MAKING MONEY

Aristotle referred to 'oikonomia' as the art of living. As part of this, the economy is aligned to nature's and society's life-giving processes. It is based on recognition of and respect for ecological limits of nature and the rights of all human beings.

He differentiated 'oikonomia' from 'chrematistics' (or the art of money-making) based on the limitless extraction of resources from nature, and wealth created by peasants, workers, and women. When chrematistics displaces oikonomia, the art of taking and grabbing displaces the art of giving. Nature becomes poorer. The ecological crisis is the poverty created for nature by not giving back to the Earth. Poverty grows as hunger, disease, homelessness, displacement, and refugees become the net results of not giving back to nature and society.

MYTH 3: MONEY IS WEALTH

Wealth used to refer to happiness and the state of well-being, the real wealth of people. It was reduced to money in recent centuries. However, in the changing world, life was reduced to money. Currency means flow. It is the flow of life and love through the web of life – in nature and society – that sustains us as one. 'The currency of life is life, not money'; that was the meaning with its roots in the Latin, which was used until the seventeenth century. It was John Locke who extended it to 'circulation of money' to suit the needs of the money-centred structures being built. The delusion that money is the currency of life has thus allowed money-making and money-makers to be rewarded and even worshipped, while our sense of interconnectedness is extinguished, and with it our potential for compassion. The fact that there is this upswing in lifestyle diseases stemming from anxiety and depression – as this money-driven life is leading to overwhelming our social, emotional, and biological systems – indicates that money and wealth may be at odds with each other.

MYTH 4: CAPITAL CREATES WEALTH. NATURE REDUCED TO LAND AND HUMANS REDUCED TO LABOUR ARE INERT, NON-CREATIVE INPUTS

Money was mutated into 'capital' as a creative force. It falsely connects creativity and creation of value to a construct named 'capital', an abstract based on money. Money is a means of exchange that reflects real value of real goods and services, created through the real work and contributions of real nature and real people. Abstracting money as 'capital' separates it from reality. The abstraction allows misplacement of creativity. It dehumanizes our understanding of the impact our choices as consumers has on people and the planet by reducing their pain and suffering into numerical values that we can easily absolve ourselves from.

The concentration of wealth and money accumulated through violence and wars was mystified into 'capital'. And with it, the creative Earth was declared dead as mere raw material, and creative, free human beings and communities were also turned into passive inputs with the label 'labour'.

MYTH 5: ECONOMIC GROWTH IS NECESSARY TO REMOVE POVERTY AND ADDRESS ECOLOGICAL PROBLEMS

Economic growth measures the extraction from nature and society. It extracts resources, minerals, soil fertility, and water from the earth. It is the cause of the creation of poverty and the destruction of nature.

'Growth', measured as money-making, is based on enclosing the Commons that sustain life and create real growth – growth in life, well-being, and happiness. Matt Bruenig, founder of the left-leaning think tank People's Policy Project, broke down the Federal Reserve's *Distributive Financial Accounts* data series and found that overall, 'the top one percent owns nearly $30 trillion of assets while the bottom half owns less than nothing, meaning they have more debts than they have assets.' The growth of wealth inequality over the last 30 years, Bruenig found, is 'eye-popping'. 'Between 1989 and 2018, the top one percent increased its total net worth by $21 trillion,' he adds. 'The bottom 50 percent actually saw its net worth decrease by $900 billion over the same period.'

BHUTAN AND THE GROSS NATIONAL HAPPINESS INDEX

Gross National Happiness (GNH), a term coined by His Majesty the Fourth King of Bhutan, Jigme Singye Wangchuck, in the 1970s, implies that sustainable development should take a holistic approach towards notions of progress and give equal importance to non-economic aspects of well-being. The GNH Index is a single-number index developed from 33 indicators categorized under nine domains with a total of 124 variables. The GNH Index is constructed based upon a robust multidimensional methodology called the Alkire-Foster method.

The key areas of GNH fall within the domains of psychological well-being, health, time use, education, culture, good governance, ecological resilience, community vitality, and living standards. Every citizen is profiled on the basis of these 124 variables, and an aggregate of 66 percent or more is taken as the indicator of happiness.

Thakur Singh Powdyel, Bhutan's former minister of education, who has been one of the most eloquent spokespeople for GNH, says, 'People always ask how you can possibly have a nation of happy people? But this is missing the point. GNH is an aspiration, a set of guiding principles through which we are navigating our path towards a sustainable and equitable society. We believe the world needs to do the same before it is too late.'

Thus, it is important to realize that it's not the data on the index that is important. The change in the parameters of looking at wealth is what makes this an experiment worth embarking on and a model to learn from for our communities and individual lives.

5 Actions to Build Sustainable, Caring, and Slow Economies

As we write this book, we are witnessing three pandemics simultaneously. The first was the Covid pandemic of 2020. The second is the pandemic of hunger. The third is the pandemic of livelihood destruction. The Covid pandemic infected 3.19 million people and killed 228,000 globally between 2020 and 2022. The World Food Programme has warned the world community of the looming 'hunger pandemic', which has the potential to engulf over a quarter of a billion people whose lives and livelihoods will be plunged into immediate danger.

There is also a pandemic of loss of livelihoods. According to the International Labour Organization (ILO), 'as a result of the economic crisis created by the pandemic, almost 1.6 billion informal economy workers (representing the most vulnerable in the labour market), out of a worldwide total of two billion and a global workforce of 3.3 billion, have suffered massive damage to their capacity to earn a living. This is due to lockdown measures and/or because they work in the hardest-hit sectors.' As pointed out by Guy Ryder, director-general, ILO: 'For millions of workers, no income means no food, no security and no future. [...] As the pandemic and the jobs crisis evolve, the need to protect the most vulnerable becomes even more urgent.'

All three pandemics have their roots in an economic model based on profits, greed, and 'extractivism', which has accelerated ecological destruction, aggravated loss of livelihoods, increased economic inequality, and polarized and divided society into the 1 percent and 99 percent.

We have an obligation to create economies that don't destroy nature, livelihoods, and the rights of workers; economies that don't destroy our health by spreading disease and pandemics; don't destroy the freedom, dignity, and the right to work; and don't create hunger. Let us build economies with the consciousness that all lives are equal, that we are part of the Earth, and we are ecological, biological beings; working is our right and is at the heart of being human; and caring for the Earth and each other is the most important work. There are no disposable or useless people. We are one humanity on one planet. Autonomy, meaning, dignity, work, freedom, and democracy are our birthrights.

Action One: Slow Money

The modern-day adage 'time is money', with images of people with a mobile phone to the ear in confident, powerful strides has become a popular culture visual for success. These people, wheeling and dealing with precise and calculated moves, look like infallible, efficient machines; they are the winners in life, and the others who are not like them are losers. This is the new version of 'happily ever after'. We have wizened up about fairy tales, having realized that real relationships begin where the 'happily ever after' ends, not to mention the problematic gender-role aspirations that these fairy tales thrust upon us. Well, now perhaps the same is required with this fast-money fairy tale – a tale so dangerous that it will cost us our bodies, our minds, and our planet. The stress and mental health issues it creates, the speed that damages our body, and the ecological impact of this lifestyle have been discussed at length. Here are four ways to break the nexus of time and money, and build futures that work for everyone:

1. Cut down the financial clutter in your life. Don't overspend, and buy only that which is important and essential for your life. Allow yourself occasional splurges and appreciate what you have. For time is free, but things cost money.

2. With the money that you saved, be an ethical investor, that is, invest in initiatives that are good not only for your financial health but for the planet and other people. Take this a step further and invest in local self-help groups in your community. Try and put 1 percent of your assets in local food systems.

3. Prevention is better than cure, so look after your body and mind. Ask yourself if the stress-induced diabetes is worth the money that you are earning? Will the carpal tunnel syndrome aid you well in your old age? Put your health first. Consumption cannot give you any joy if your body and mind don't cooperate with you.

4. Invest as collectives of friends and family towards the future you want for the next generation – build relationships with the Earth, communities, your friends and family, and life itself.

SLOW MONEY MOVEMENT

Woody Tasch, socially conscious investing pioneer and author, coined the word and movement 'Slow Money'. Borrowing the conceptual framework of 'Slow Food' – wherein participants reject convenience-oriented 'fast' foods and instead fill up their plates with traditional, unprocessed and, ideally, locally produced foods – Tasch applied the same principles to personal finance and investing. Thus, Slow Money is dedicated to connecting investors to their local economies by encouraging financial resources to be invested in small food enterprises and local food systems. Instead of venture capital bankrolling far-flung high-tech start-ups, Tasch hopes to see 'nurture capital' funding local merchants and producers who in turn plug half of their profits back into their communities, ensuring one small local virtuous circle that values soil fertility, carrying capacity, a sense of place, care of the Commons, diversity, non-violence, and cultural, ecological, and economic health as much as financial return.

Action Two: Bread Labour – Enabling Farmers and Community Markets

'As famine struck with unparalleled ferocity, scores of farmers starved to death, and their children, with blobs of snot seeping down their noses, with their bellies distended and their bodies horribly mangled, the children looked at the bleak horizon with the sort of beseeching eyes one prays no one gets to lock into a stare. The villagers were dying; there was no money, no food, no water, and no hope. Forced to cultivate cash-crops by the ruthless State and the unforgiving landlords, the farmers realized to their collective horror that one can only sell a cash-crop, not consume it. But what happens when those buying the produce have fixed the price to their liking? The only way out is suicide.'

P. SAINATH

Hyper concentration of corporate power and convergence of technologies of digitalization, financialization, and commodification for total control over daily bread typifies the next step of the drive for profits. Digital Finance or FinTech (the marriage of finance and technology) is putting money and the money machine in the driver's seat to decide whether farmers grow real food for real

people, or commodities for Big Retail, and whether we eat real food or fake food packaged as 'natural'.

'Financialization of ecological services' is the new vocabulary for the food and agriculture paradigm, separated from the social, ecological, ontological, and epistemological fabric of farming.

The concentration of corporate control over our daily bread through large mergers of giant seed and chemical corporations, as well as the convergence of tools of digital and financial technologies, along with the integration of the entire food chain from the seed to our table through these interconnected technologies, implies higher vulnerability of the food system as a whole. An instability in one part can trigger instabilities throughout. Though the instabilities begin in the finance world, they translate into the vulnerability of real people and real lives. When food and agriculture are put in a financial casino, the collapse of the financial systems immediately translates into a collapse of the food system. The planetary emergency, and the social and economic emergency of unemployment, farmers' debt, and suicides are thus interconnected, and are a creation of the fast economy logic. A major study shows that only 1 percent of the consumer price reaches the farmer in the supply chains controlled by large corporations. The creation of local ecological economies, based on 'bread labour' and co-creativity with nature, is the only way to sustain the Earth and human societies.

FAIRTRADE – LOOK FOR THE MARK

The majority of our food is produced by small-holder farmers. However, due to systemic challenges like lack of access to finance, infrastructure, and markets, many are unable to meet their daily needs. Fairtrade is an alternative approach to trade that prioritizes a better income for farmers and workers, and taking care of the planet. Fairtrade changes the way trade works through better prices, decent working conditions, local sustainability, and fairer terms of trade for farmers and workers in developing countries. Fairtrade aims to tackle the injustices of conventional trade, which traditionally discriminates against the most vulnerable. When you see the 'FAIRTRADE' mark on food (fruits, coffee, tea, chocolate, and spices, among others) and fashion products, it means that the product has been made following a system of standards and certification that ensures a better deal for the producers in the supply chain, and that better environmental standards are being followed. Fairtrade International is a network of not-for-profit organizations that drives the strategy for bringing Fairtrade into the mainstream and to represent the best interest of the farmers and workers behind our food and fashion choices. It is one of the few global networks where farmer and worker organizations have a 50 percent say in the decision-making. Working with a wide range of stakeholders, including farmers, trade unions, academicians, consumer organizations, education institutions, large and small businesses, Fairtrade works towards a more sustainable future for all and incorporates the United Nations Sustainable Development Goals (UNSDG) in our food and fashion choices.

To learn more about the movement, visit www.fairtrade.net

Action Three: Sacred Economy

EMPOWERING ARTISANS AND HANDICRAFTS, RECONNECTING WITH THE BEAUTY OF EVERYDAY THINGS

'The growth of crafts in society was the sign of the cultivation of sensitivity and the stirring and mellowing of humanism.... Here we see the transformation of the mere functional into works of art.... No aspect of life was too insignificant or humble to lay claim to beauty or acquire sanctity as a symbol of good omen.... One may say that there is no such thing as craft for craft's sake. It is not an exercise of the intellect, which finds a form in a secluded studio to become the proud possession of an individual or institution. It arises from the deeper hunger of humanity, its functions are socialized for its use and distributed through the family and even the entire community.'
KAMALADEVI CHATTOPADHYAY
UNESCO's publication, The Arts and Man (Paris, 1969)

The Earth – life in its rich diversity – is sacred. Seed and food are sacred. Land and soil are sacred. Water is sacred. Our rivers

and forests are sacred. Our bodies are sacred. Creative sovereign work in service and care for the Earth and community is sacred. According to Prasanna Heggodu, founder of Gram Seva Sangh, 'Sacred sectors are sectors with a minimum of 60 percent human labour and 60 percent local raw materials, and less than 40 percent use of automation and less than 40 percent imported raw materials.' Thus, the handicraft industry is the exemplar of the sacred economy. It creates jobs for a large population of the Earth, and work done by hand takes less energy than a mass production assembly line, which makes it environmentally more sustainable, thereby addressing the current climate crisis. Making local economies stronger, it creates a better social environment without strife and reduces pressure on cities, as it revives villages and reduces migration to urban areas.

In a growing technology-run and industrialized world, people are increasingly forgetting traditional crafts and moving towards innovations. Due to this, artisans are moving to alternative income-generation methods and often taking up daily-wage labour work to sustain themselves. With this shift, preserving, protecting the skills, and knowledge of traditional crafts is a growing challenge.

In the era of plastic and synthetic materials, we are moving away from inviting beauty into our everyday lives. The notion of monoculture beauty makes us look at the uniqueness of handicrafts and crafts as imperfections and flaws. Things that our grandparents used to use in their everyday lives – beautiful nutcrackers, carved wooden combs, pottery – have now made it to our living room display cabinets. When did we get divorced from beauty in our everyday life for the sake of mere functionality?

In the history of Western aesthetics, the subject matters that received attention ranged from natural objects and phenomena, built structures, utilitarian objects, and human actions, to what is today regarded as the fine arts. However, from the nineteenth century onwards, the discourse has become increasingly focused on the fine arts. Challenges to this rather limited scope of aesthetics began during the latter half of the twentieth century with a renewed interest in nature and environment, followed by the exploration of popular arts. Everyday aesthetics continues this trajectory of widening scope by including objects, events, and activities that constitute people's daily lives. However, it is more accurate to characterize this recent development as *restoring* the scope of aesthetics rather than opening a new arena.

Over the last 50 years in particular, chronic excess has led to a world of big business, globalization, overpopulation, and all the destructive attributes these create: pollution, greed, climate change, and mass inequality, among others. Common things in commonplace settings tend to get overlooked in the grand scheme of wants and desires. You don't need a helicopter, a Ferrari, or even a mansion to experience the best of life. In an age of feeble and ugly machine-made things, we need to deepen and transform our relationship with the objects that surround us.

The philosopher and founder of Japan's *mingei* (folk craft) movement, Soetsu Yanagi believed in the aesthetic appreciation of the everyday things that surround us. As common things in commonplace settings tend to get overlooked in the grand scheme of wants and desires, Yanagi asks us to shift our focus to a natural, sturdy, and safe aesthetic – objects that can last a lifetime, the exact opposite of our gadget-crazy lust for the latest device. Shifting our mind's eye to everyday things and witnessing the beauty in them is an act of inviting love into our lives. As he says, 'Even the common articles made for daily use become endowed with beauty when they are loved.' Invite love and beauty back into your life and nurture sacred economies by:

1. Buying natural products.

2. Buying handmade, handcrafted products.

3. Learning about the history, culture, and the people who made the everyday things you use – plates, bowls, flower pots, clothes, shoes, and storage units – so that they stop becoming just objects you purchased but stories of care and love instead.

4. Work with craftspeople and artisans to make things that work

for you and your home; play with material, play with design, experiment, and co-create.

5. Slow down and enjoy the beauty of everyday things; notice them while eating out of them or using them; discover them through observation and touch. Do everyday things slowly – eat slowly, clean slowly, and dress slowly so you may enjoy the beauty in them all.

6. Learn about the local handicrafts of your area; teach your children about various crafts, the texture of textiles and designs. Learn the stories and mythologies that surround them so that when you buy something handmade, you are connected to a story that has travelled many generations and will travel many more.

7. Most artisans struggle with having access to an e-portal for their products, which means local exhibitions and melas are their best chance of getting market access. Make an effort to visit such exhibitions and get in touch with the artisans. Procure directly from them whenever you can down the line, especially for gifting and festivities.

8. When buying on big e-portals, check the names of brands and then try and locate the original vendor. This will help you in two ways: one, you will bypass the 'middle-man', and second, you can then request the local vendor to pack and send you the product plastic-free – something several big e-portals do not do.

9. Mend and upcycle things as much as possible. A lot of local handicraft was about maintaining and transforming existing things, so add a little embroidery to old clothes to make them new, transform old furniture with embellishments, and get the local cobbler to fix your shoes. There are many ways that you can add a little flair by upcycling or mending your things. Building a relationship with craftspeople is key to sustaining the crafts, and mending builds long-term relationships.

Action Four: Supporting Local Hawkers

The giant retail chains have their own standards of buying produce from farmers and producers. Without using excessive pesticides, it is very difficult for a farmer to produce fruits and vegetables that fit into these standards. Once these farm products come to the retail giants, they sell them throughout the year by preserving them in cold storage; but in the process, a number of preservatives are also added to the food. When a consumer buys a 'preserved' vegetable from these stores, it is full of toxic materials that are harmful for consumption.

Climate change due to air pollution is already becoming a threat to human life. Temperatures are rising, the sea level is rising, and glaciers are melting. The imperative in the context of climate change is to prevent an increase in the use of fossil fuels. Our hawker, *raddiwala*, and kirana stores are the solution to climate change.

Models like the Reliance Store and Bharti-Walmart increase the fossil-fuel use and carbon emissions by increasing the food miles. Local hawkers are under the threat of displacement by the authorities, and their business is eroding due to corporate e-commerce by large corporations like Walmart and Amazon. According to various reports, in India, their revenues have reduced to the extent of 40 percent and 20 percent for hawkers and small traders. The short chains reduce the food miles while enhancing the freshness, quality, and rich biodiversity of food on your table.

Action Five: Slow Work, Slow Offices, and You

On average, we spend eight hours or more at workplaces, for we must work to pay the bills. If we are lucky, we may enjoy what we do for a living. More than half of our adult life is spent in an office space, thus we must consider its ecological impact to be part of our work culture. From the way we design our office spaces to the social culture within the workplace, to the everyday requirements of running the office, we can gradually shift to an ecological perspective. Here are some baby steps we can take to make that shift:

1. CHOICE OF STATIONERY AND OFFICE SUPPLIES

i. Buy sustainable stationery made from biodegradable materials. Many cheap options are currently available in the market.

ii. When printing, use both sides of the paper. If you must print on one side, make notebooks or pads for rough work from the blank side.

iii. Go paperless. Wherever possible, print less in general; ask vendors to offer e-statements and invoices instead.

2. COFFEE AND TEA BREAKS

i. Don't use single-serve packaging cups; use your own mug to reduce the creation of garbage.

ii. Serve organic and fair-trade coffee and tea.

iii. Make your coffee break station or room fun and attractive so that more people would choose to have their break in the office rather than head to a coffee shop.

iv. Tea bags and coffee filters can be useful, but are mostly unnecessary. Great coffee can be made at home with a reusable filter or a stovetop espresso maker. A quality tea infuser can last a lifetime and replace an untold number of (questionably compostable) tea bags. If you do use filters and bags, look for biodegradable and unbleached ones.

3. CLEANING AND MAINTENANCE

i. Switch to herbal, eco-friendly cleaning chemicals.

ii. Instead of using floor carpets and hard-to-clean furniture, consider installing tiles and using wood wherever possible. Not only are they much easier to clean and can usually be kept looking nice with a simple wipe-over or mop, it also results in less usage of toxic chemicals.

4. ENERGY CONSUMPTION

During breaks, turn off all lights, shut down your computers and other appliances, and ensure that all windows are closed.

5. TRANSPORTATION

i. Car pool to work.

ii. If your job permits, work from home.

6. GREEN CONFERENCES AND MEETINGS

i. Consider making small switches like recyclable name badges, pens, notepads, and paperless agendas.

ii. The single most impactful choice you can make for a greener event is where you host it. Choose places that have sustainable practices or request your venue partners to adopt a few. Starting with not having bottled water at your events to choosing places that have adopted water harvesting and are supportive of local businesses. How far from the airport is the convention centre? How far apart are the convention centre and your recommended hotels? The closer together, the shorter the ride to the convention centre, which typically means a lower carbon footprint. Consider the walkability of the convention centre district. How many hotel rooms are located within walking distance (less than 1 kilometre) of the convention centre? What about restaurants? The more accessible the main venue is to attendees on foot, the lower the environmental impacts.

iii. To reduce waste, bring fewer items on site. Encourage exhibitors to reduce or eliminate on site giveaways that often end up in the trash. Track what is left over at the end of your events. If you can reduce the quantity to be produced, you can save money and save space in the landfill.

iv. Opt for event food and beverages that are sustainable. Incorporate as many seasonal/local and organic food offerings as possible, and ask if the venue has an on site garden or local or regional food vendors that can round-out your menus.

v. Recycle and donate unused materials. Things like vinyl signs and poster boards could be repurposed. Look for organizations that make it easy for attendees to recycle – place recycling bins next to trash cans so guests can easily deposit recyclables.

HARVEST FESTIVALS ARE CELEBRATED IN MOST CULTURES across the world, from Thanksgiving in the United States, to Makar Sankranti and Onam in India. Laced with cultural stories of overcoming hardship and feasting, both family and public use foods drawn from crops that come to maturity around the time of the festival.

Onam is a harvest festival celebrated in the first month of the Malayali calendar. It commemorates the mythical ruler Mahabali, a beloved king who defeated the Hindu gods in battle only to be banished to the underworld by Vishnu in his Vamana avatar. In recognition of his just and prosperous reign, Vishnu allowed Mahabali to return to his people once a year. People across Kerala celebrate both the harvest and Mahabali's return by making ornate flowered *pookalam* designs, dancing the *Pulikali* and *Kummattikali*, and competing in *Vallam Kali* or boat races. The most delicious tradition, however, comes on *Thiruvonam*, the climax of the festival, in the form of the Onam sadhya, meaning 'Onam feast' or 'banquet' in Malayalam.

READING LIST

VANDANA SHIVA, *Oneness vs. the 1%: Shattering Illusions, Seeding Freedom.* Spinifex Press (2018).

ELA R. BHATT, *Anubandh: Building Hundred-mile Communities.* Navajivan Trust; 1st edition (2015).

M.K. GANDHI, *Hind Swaraj and Other Writings.* Cambridge University Press (1997).

WOODY TASCH, *Inquiries into the Nature of Slow Money: Investing as if Food, Farms, and Fertility Mattered.* Chelsea Green Publishing (2010).

YANIS VARDOULAKIS, *Technofeudalism: What Killed Capitalism.* Vintage Publishing (2023).

Recipes to Celebrate Slow Economies

KHULAD PHIRNI

INGREDIENTS

3 tbsp rice
1 tsp ghee
1 litre full-fat milk
15–20 strands of saffron
½ cup sugar
½ tsp rose water or green cardamom powder
Almond slivers for garnishing
Earthen bowls (khulad) for serving

METHOD

- Wash the rice in running water and soak for an hour.
- Drain the water. In a blender or mixer-grinder, blend the soaked rice to a coarse paste. You can add a little water, if required.
- In a heavy-bottom pan over medium heat, add the ghee. Once the ghee is hot, stir in the milk.
- Add the prepared rice paste and saffron. Let the milk simmer until it is reduced to half, and the phirni has reached the desired consistency. It will further thicken on cooling.
- Add the sugar and cardamom/rose water. Remove from the heat and let cool.
- Transfer the phirni into earthen bowls. Garnish with almonds. Refrigerate for a few hours before serving.

Long before we started using metal utensils, our forefathers have used utensils and storage items made of mud. Some of these were reusable along with being completely disposable and eco-friendly.

KOLKATA'S *street food* DELIGHT
JHAL MURI

INGREDIENTS

250 gm puffed rice
2 tbsp finely chopped onion
2 tbsp boiled and mashed potato
2 tbsp finely chopped tomatoes
2 tbsp finely chopped cucumber
2 tsp finely chopped ginger
4 tsp finely chopped green chillies
2 tbsp peanuts, fried or roasted
2 tsp sprouted black gram soaked, drained
4 tsp mustard oil
Salt, to taste

METHOD

- Leaving out the mustard oil and salt, gently combine all the ingredients in a large bowl.
- Once done, add the mustard oil and salt. Mix again to combine. Serve as a snack.

In Bangla, 'jhal' means spicy and 'muri' means puffed rice. A special ingredient that is added in Jhal Muri is mustard oil. Source the ingredients from your local vendors to support the local economy.

CELEBRATE ABUNDANCE *with* AVIAL ONAM SADHYA

INGREDIENTS

1 raw banana
½ cup chopped yam
2 drumsticks
¼ cup cluster beans
¼ cup snake gourd
1 carrot
1 cucumber
1 cup grated coconut
5 green chillies
½ tsp cumin seeds
½ tsp turmeric powder
Salt, to taste
½ cup beaten curd or
5 tbsp raw mango
1 sprig curry leaves
1 tbsp coconut oil

METHOD

- Peel and cut all the vegetables lengthwise into 2-inch-long pieces. Wash and keep aside.
- In a mixer-grinder, grind the grated coconut with green chillies and cumin seeds to a coarse paste. Use a little water if the paste is too thick.
- In a pot over medium heat, add half a cup of water, turmeric powder, and season to taste with salt. Add the vegetables to the boiling water according to the cooking time – the longer cooking vegetables first and the quicker ones last. Cook until they are half cooked.
- Make a small well in the middle of the half-cooked vegetables and pour the ground coconut paste into it. Cover the pot and let simmer for 5 minutes. Remove the lid and add the beaten curd and curry leaves. Gently toss for few minutes. If you are using raw mango, don't add the curd.
- Cook over low heat until all the vegetables are done. Take care not to mash them. Add the curry leaves and pour the coconut oil over the top and cover the pot. Remove from heat and keep covered until it is time to serve.

(Source: https://www.kothiyavunu.com/2009/09/aviyal/)

9

SLOW DEMOCRACY

We are of the Earth, in our diversity. The Earth gives us citizenship.

As human beings, our first identity is as Earth Citizens, an Earth Family (*Vasudhaiva Kutumbakam*), sharing the planet with other species. We did not impose on ourselves the burden of anthropocentrism, of separation from and superiority over other species. The colonial empire was based on the illusion of the superiority of one species, one race, one religion, and one gender. It was also an empire over what were declared to be 'lesser creatures', to be exploited and exterminated. The monoculture of the mind is the basis of an empire.

We are part of *Vasudhaiva Kutumbakam*. As members of the Earth family, we have natural rights that flow from being part of nature. The fact that nature has rights and the fact that the discourse on 'rights of nature' has taken flight are positive developments. The future demands that our concepts of 'political', 'economic', 'social', and 'human rights' reflect our being part of this 'one world' which, on one hand, creates responsibilities and duty to care and share, and on the other, creates our natural rights to life, to food and water, and to air and space, which we share with other beings – the Commons that represent the true meaning of democratic life.

Large institutions, centralized governments, and top-down thinking are no longer society's drivers. New decision-making techniques are ensuring that local communities and their citizens are uniquely suited to meet today's challenges. Connecting in a meaningful way with our community and reclaiming our power as citizens is both powerful and possible. We have to recognize that representative democracy must be rooted in the fertile soil of face-to-face, local, problem-solving democracy.

Citizen advocacy at the local level brings to mind the story of David and Goliath. When communities get together against corporations and ban genetically modified crops or win back the right to create a local energy plan or market local foods, some feel hope for an empowered future. For some, however, local advocacy is a source of pain and scepticism. Get real, they say; in the long run, the big money behind projects like genetically modified foods and fracking for natural gas will always win, no matter how earnest the local efforts.

If we want to re-energize democracy, it has to be more than the same old political system transposed to the community level. Instead, we need slow democracy, which is not a clarion call for endless meetings but a reminder of the care that's needed for full-blooded community decision-making.

Slow democracy weaves together three key elements of democratic decision-making: inclusion – ensuring broad, diverse public participation; deliberation – defining problems, weighing options, and co-creating solutions through sound information and respectful relationships; and power – defining a clear connection between citizen participation, public decisions, and action.

We desperately need leaders with clarity, wisdom, and courage – and voters to elect them. If people have held leadership positions themselves, they will have a much better understanding of what real political leadership entails. They will know not to expect agreements overnight and be able to distinguish between bullying and respectful deliberation as well as between political posturing and authentic debate for consensus-building.

Fast Democracy vs. Slow Democracy

FAST DEMOCRACY	SLOW DEMOCRACY
Country-based patriotism and allegiance to one nation; divisive and segregated.	**Earth citizenship – interconnections, and respect and celebration of diversity.**
Centralized governance – a one-size-fits-all solution done to scale, disregarding contextual conditions.	**Decentralized governance based on contextual conditions and localized context, creating a fractal of change.**
Mile-wide, inch-deep.	**Inch-wide, mile-deep.**
Individual rights and political entitlement.	**The rights of Mother Earth and Universal Responsibility.**

SLOW CITIZENSHIP: 10 PRINCIPLES OF EARTH DEMOCRACY

1. ECOLOGICAL DEMOCRACY – DEMOCRACY OF ALL LIFE

As members of the Earth, we have the duty to protect the rights and welfare of all species and all people. No humans have the right to encroach on the ecological space of other species and people, or treat them with cruelty and violence.

2. INTRINSIC WORTH OF ALL SPECIES AND PEOPLES

All species, humans, and cultures have intrinsic worth. They are subjects, not objects of manipulation or ownership. No humans have the right to own other species, people, or the knowledge of other cultures through patents and other intellectual property rights.

3. DIVERSITY IN NATURE AND CULTURE

Defending biological and cultural diversity is a duty of all people. Diversity is an end in itself – a value, a source of richness – material, intellectual, emotional, and cultural.

4. NATURAL RIGHTS TO SUSTENANCE

All members of Earth, including humans, have the right to sustenance – to food and water, to safe and clean habitat, to security of ecological space. These rights are natural rights; they are birthrights given by the fact of existence on Earth, and are best protected through community rights and Commons. They are not given by states or corporations, nor can they be extinguished by state or corporate action. No state or corporation has the right to erode or undermine these natural rights or enclose the Commons that sustain all through privatization or monopoly control.

5. EARTH ECONOMY IS BASED ON ECONOMIC DEMOCRACY AND LIVING ECONOMY

Earth democracy is based on economic democracy. Economic systems in Earth democracy protect ecosystems and their integrity; they protect people's livelihoods; and provide basic needs to all. In the Earth economy there are no disposable or dispensable species or people. The Earth economy is a living economy. It is based on sustainable, diverse, pluralistic systems that protect nature and people, are chosen by people, for the benefit of the common good.

6. LIVING ECONOMIES ARE BUILT ON LOCAL ECONOMIES

Conservation of the Earth's resources and creation of sustainable and satisfying livelihoods are most caringly, creatively, efficiently, and equitably achieved at the local level. Localization of economics is a social and ecological imperative. Only goods and services that cannot be produced locally, using local resources and local knowledge, should be produced non-locally and traded over long distances. Earth democracy is based on vibrant, resilient local economies, which support national and global economies. Where we do not view our existence as merely consumers that celebrate having access to products from every corner of the world, without reflecting on how such consumption impacts climate and undermines diversity through monocultures of food, clothes, and beauty standards. We need to be mindful that the way we spend money impacts the rights of Mother Earth and all its inhabitants – what is commonly termed as 'voting with your wallet'.

7. LIVING DEMOCRACY

Earth democracy is based on local living democracy, wherein local communities – organized on principles of inclusion and diversity with ecological and social responsibility – have the highest authority on decisions related to the environment, natural resources, and the sustenance and livelihood of people. Authority is delegated to more distant levels of governance on the principle of subsidiarity.

8. LIVING KNOWLEDGE

Earth democracy is based on earth-centred and community-centred knowledge systems. Living knowledge is knowledge that maintains and renews living processes and contributes to the health of the planet and people. It is also living knowledge in that it is embedded in nature and society, is not abstract, reductionist, and anti-life. Living knowledge is a Commons; it belongs collectively to communities that create it and keep it alive. All humans have a duty to share knowledge. No person or corporation has a right to enclose, monopolize, patent, or exclusively own as intellectual property living knowledge. Personal data and hyper-surveillance is questioned in such a system, and technology is people-centric and not a tool for corporate and political exploitation.

9. BALANCING RIGHTS WITH RESPONSIBILITY

In Earth democracy, rights are derived from and balanced with responsibility. Those who bear the consequences of decisions and actions are the decision makers.

10. GLOBALIZING PEACE, CARE, AND COMPASSION

Earth democracy connects people in circles of care, cooperation, and compassion instead of dividing them through competition and conflict. Earth democracy globalizes compassion, not greed, and peace, not war.

Learning from Indigenous Communities

'Even if you keep telling/calling us uncivilized, primitive, backward or whatever for humiliating us, we're not going to handover our remaining natural resources to fulfil your greed.... We're neither backward nor uncivilized. Our knowledge is the best for the sustainable future of the Planet.... We're not selfish. We're not just fighting to protect our land, territory and resources. But we're also fighting to protect your future..... The future of the planet depends on the fight of the Indigenous Peoples. If we give up, the greedy corporate in nexus with the States, will swallow the remaining natural resources.'

GLADSON DUNGDUNG

Adivasis and Their Forest

The Western paradigm of natural rights as involving the inalienable right to property, is therefore, completely flawed. This is why it is the 'tyranny' of rights which is at the root of all the violence, terror, rape, exploitation, and deprivation today. Until we stay in this paradigm of rights, this violence and destruction will only grow. By telling us that rights don't just flow from the fact of being an ecological and biological being, what this Western

paradigm has ingrained in us is the idea that being who we are, our very existence, is not enough. And this has inflicted massive, pervasive and deep mental, emotional, and spiritual damage on all of us.

If this rights regime is not corrected, we will be led on a path as devastating and destructive as our colonial pasts have been. And this time, colonization will end in the extinction of biodiversity, including the human species. This colonizing past has decimated our agricultural system, our economic system, our educational system, our political system, and more importantly, our interpersonal ways of being. The right to possess and hold property has deeply damaged and altered something inside us. It has led to a sense of xenophobic entitlement leaning into hate propaganda based on the logic of colonial empire, where some lives matter more than others. This logic creates a chasm where possession, rather than trusteeship, drives our core ambitions in life. Dismantling communities and making our human journey more individualistic breaks our interconnection with other living beings, and hyper-privatization robs us of any sense of belonging and worth derived from acknowledging that we are part of the Earth family.

Earth democracy, then, is the potential for a shared future – one where we all recognize and celebrate the natural rights arising from being a part of the Earth family; where these rights are born from the ecology of our being; where the very membership of every species and living creatures of one Earth implies the existence of these natural rights.

In the figure on the following page,[1] Quadrants A and B represent the Western paradigm of rights, which is rooted in viewing all of us as individuals and a set of individuals with no permeability. Its spectrum remains anthropocentric and restricted to human beings. If we remain in that paradigm, the seeming 'expansion' of rights will still be restricted, as it will still see everything as separate with distinct boundaries.

On the other hand, Quadrants C and D represent the paradigm of the rights of the Earth family. It perceives every living being in relatedness with every other living being. There is a permeability which is recognized. Each being is also seen as carrying the past and future time and space with it and within it.

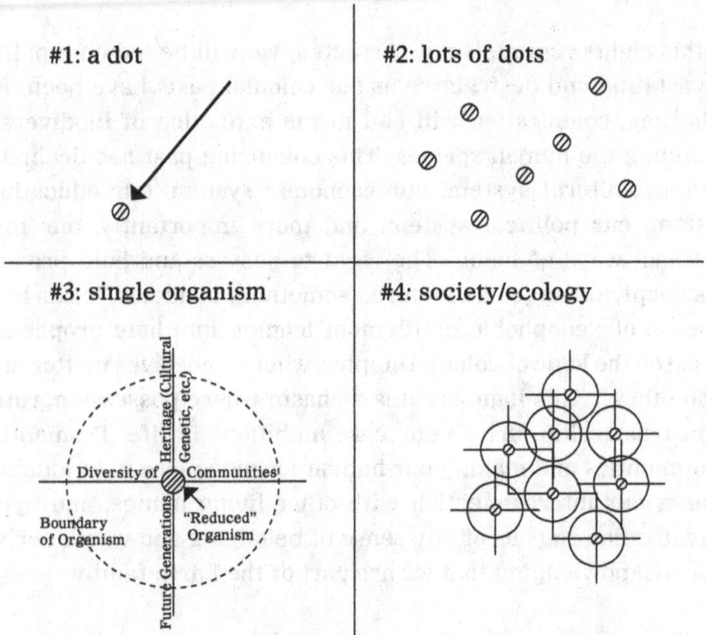

Thus, democracy as a manifestation of ecology and nature has to be a system of fractals of ideas and practices. As Adrienne Maree Brown says in her book *Emergent Strategy*, 'we must be like the oak trees where they hold on together sub-terraineally through a network of strong roots to weather a storm but stand tall on the surface; we must be like mushrooms converting toxicity into beauty and sustenance and be like the dandelion, a community of healers who spread far and wide for one flower is enough to create a field of it and the health benefits of dandelion is immense.' As we create these communities which respect diversity and interconnectedness a truly healthy democracy can thrive in it.

---

1. This framework is built upon what Brewster Kneen has depicted in *The Tyranny of Rights*. Ottawa: The Ram's Horn, 2009.

LANDMARK COURT CASES AND LEGAL RECOGNITION

The landmark **Mabo Case (Mabo vs. Queensland (No. 2)**[1] was a decision in the Australian Courts in 1992, which gave native title in Australia for the first time. 'Native title' was described as 'the interests and rights of indigenous inhabitants in land, whether communal, group, or individual, possessed under the traditional laws acknowledged by and the traditional customs observed by the indigenous inhabitants.' This case pertained to the rights and interests of the Meriam people to their traditional land.

The main question that was framed in the case was whether the rights of the indigenous people survived after the colonization by the Crown. Justice Brennan, in para 61 of the judgement, answered: '*Mere change in sovereignty does not extinguish native title to land.*'

He quoted another case called **Calder vs. Attorney-General of British Columbia**,[2] where Justice Hall had rejected as 'wholly wrong' 'the proposition that after conquest or discovery the native peoples have no rights at all except those subsequently granted or recognized by the conqueror or discoverer'. Similarly, in India, we have the **Fifth Schedule of India's Constitution**, along with many state laws, which affirms the non-alienability of tribal land as an elementary right.

These examples reflect the idea that natural rights are ontological, ecological, and biological. These ontological rights have always existed and will continue to exist. No human law or court can 'create' these rights. It can only affirm them. They exist de facto.

A statement from 2003 from a meeting of indigenous people in the autonomous communities in Mexico states:

> ... the government resolved not to recognize our fundamental rights in the constitution, but to intensify its plundering, destruction and robbery policies towards our lands, territories, and natural resources... confronted with the aforementioned, we have decided to stop demanding further recognition for the exercise of our own rights, so now we demand respect for our lands, territories, and autonomy. We have resolved that this state has lost its legitimacy by its legal practices; we must exercise our autonomy de facto, thus addressing our grave situation and looking forward to a better future for our children.[3]

---

1. Mabo vs. Queensland (No. 2) HCA 23, (1992) 175 CLR 1.
2. Calder vs. Attorney-General of British Columbia (124) (1973) SCR.
3. Statement of the Pacific-Center Region of the Congreso Nacional Indigena meeting in Tlanixco state, Mexico, 25-26/1/03.

4 Actions to Build Slow Democracy

Democracy is as much a value and a philosophy as it is practice and action. So how do we convert these ideas into action in our own lives and the lives of our community? Where do we start to grow and nurture democracy? Here are four ways or frameworks in which you can practise it in your daily lives. Hopefully, they will inspire your own initiatives and practices based on your local realities.

ACTION ONE: FROM A MILE-WIDE INCH-DEEP DEMOCRACY TO INCH-WIDE MILE-DEEP DEMOCRACY – CREATE A JAIV PANCHAYAT

'Trusting that those roots go deep, trusting that those roots find nourishing soil, trusting that those roots intertwine with other roots, finding connection to the people planted in other inches... (and intersecting) (and providing collective sustenance).'

RENEE ROEDERER

Ordained Presbyterian (PCUSA) minister

Adrienne Marie Brown, on page 20 of her book *Emergent Strategy: Shaping Change, Changing Worlds,* states, 'We need each other. I love the idea of shifting from "mile-wide inch-deep" movements to "inch-wide mile-deep" movements that schism the existing paradigm.' Navdanya has been practising this for decades, and one of the most successful experiments in this for us has been the Jaiv Panchayat that we invite you in this section to explore and create for your own communities.

The 'Jaiv Panchayat' or the Biodiversity Panchayat is an example of living democracy – both in being the democracy of all life and democracy in everyday life. It consists of the entire Gram Sabha or Residents' Welfare Associations in the city, including women, children, and minority communities, and not merely those who are on the electoral rolls of the village. This form of the panchayat renders the community the decision-maker on all matters pertaining to biodiversity and its conservation. In doing so, the Jaiv Panchayat lays down the parameters within which the elected panchayat body can take action vis-à-vis biodiversity.

The first Jaiv Panchayat was brought to life by a gathering of about 1,000 villagers of Agastyamuni village in district Rudraprayag, Garhwal, Uttaranchal (now Uttarakhand) on 5 June 1999 – World Environment Day. As a part of the movement, over 4,000 village communities affirmed their rights to their biodiversity and took a pledge to conserve, rejuvenate, and protect their biodiversity. There are more than 85 Jaiv Panchayats in Garhwal alone, where people have asserted their inalienable and common rights to their natural resources. In many of these Jaiv Panchayats, the elected leaders are also the leaders of the movement. Many of them have declared their villages GM-free zones as well.

Jaiv Panchayat records the biodiversity of the village in their own Community Biodiversity Register (CBR) to protect and reclaim the biological and intellectual commons. It has rejuvenated indigenous knowledge and promoted its propagation from grandmother to grandchildren.

You, too, can help bring to life a Jaiv Panchayat and make real the idea of a living democracy – a democracy for life in all its diversity. These are some of the things you can do:

- Organize the people of your village or locality into a Jaiv Panchayat and help them understand how it would be a decision-making body on all matters pertaining to the conservation, management, and protection of all biological resources of that area.

- Organize meetings/awareness campaigns with the Jaiv Panchayat, and discuss the diverse kinds of biological wealth available and used in your area.

- Make a formal declaration that all the biological resources belong only to the community.

- Prepare a Community Biodiversity Register (CBR) to prevent erosion of biological resources and knowledge.

- A few active members (Jaiv Rakshaks) from amongst the group can take the responsibility of maintaining and updating the register periodically.

Action Two: 200 Present

RATIFYING THE RIGHTS OF MOTHER EARTH IN YOUR COMMUNITIES AT A RESIDENTIAL COMMUNITY

'A favourite concept of mine is the 200-year present, a way of thinking about change. The 200-year present began 100 years ago with the year of birth of the people who have reached their hundredth birthday today. The other boundary of the 200-year present, 100 years from now, is the hundredth birthday of the babies born today. If you take that span, you and I will have had contact with a lot of people from different parts of that span. So think in terms of events over that span and realize how long change takes. You can see how difficult it has been to create these bodies and new ways and how in many ways we are slipping backward; but in other ways we are not. I take comfort to know that super-power hegemony has a very limited lifespan (decline and fall of Rome, the Ottoman Empire).'

ELISE M. BOULDING

during an interview to Julian Portilla (2003)

Elise Boulding suggests we calculate 'the present' by subtracting the date of birth of the oldest person we have known in our lives from the projected passing-on date of the youngest person in our family. This is our relationship with history and our understanding of the past, our living of the present, and our relationship with the future. In Hinduism, one talks of praying for blessings for our seven generations to come. All these notions link our actions today to the future we are creating for the planet and its people. Thus, we must not think of ourselves to be weak, and in front of large leviathans, we must harness the impact of our time on this planet and sow the seeds of peace, justice, and freedom for all species.

By introducing to our communities the **Charter of the Rights of Mother Earth,** we engage them in a dialogue that will lead us as a collective to the awareness of how aligned our everyday living is towards caring for our coming generations.

Declaration of Rights of Mother Earth: *Article 2 of the declaration lists all the inherent rights of Mother Earth:*

1. MOTHER EARTH AND ALL BEINGS OF WHICH SHE IS COMPOSED HAVE THE FOLLOWING INHERENT RIGHTS:

i. The right to life and to exist;

ii. The right to be respected;

iii. The right to regenerate its biocapacity and to continue its vital cycles and processes free from human disruptions;

iv. The right to maintain its identity and integrity as a distinct, self-regulating, and interrelated being;

v. The right to water as a source of life;

vi. The right to clean air;

vii. The right to integral health;

viii. The right to be free from contamination, pollution, and toxic or radioactive waste;

ix. The right to not have its genetic structure modified or disrupted in a manner that threatens its integrity or vital and healthy functioning;

x. The right to full and prompt restoration of the violation of the rights recognized in this Declaration caused by human activities.

2. EACH BEING HAS THE RIGHT TO A PLACE AND TO PLAY ITS ROLE ON MOTHER EARTH FOR HER HARMONIOUS FUNCTIONING.

Every being has the right to well-being and to live free from torture or cruel treatment by human beings. Examining these rights as a community will lead to creating our actions and initiatives to align ourselves to them, such as:

i. Rainwater harvesting

ii. Community kitchen gardening

iii. Community composting

iv. Planting trees and looking after them

The innovation will flow with the rights of Mother Earth being our guiding star, and dialogue and discussion will enrich and further nuance our actions.

THE BOLIVIAN CONSTITUTION IN ARTICLE 7 (RIGHTS OF MOTHER EARTH) STATES

I. Mother Earth has the following rights:

- **To life**: To maintain the integrity of living systems and natural processes that sustain them, and capacities and conditions for regeneration.

- **To the diversity of life**: It is the right to preservation of differentiation and variety of beings that make up Mother Earth, without being genetically altered or structurally modified in an artificial way that threatens their existence, functioning, or future potential.

- **To water**: The right to preserve the functionality of the water cycle, its existence in the quantity and quality needed to sustain living systems, and protection from pollution for the reproduction of the life of Earth and all its components.

- **To clean air**: To preserve the quality and composition of air for sustaining living systems and its protection from pollution. For the reproduction of the life of Mother Earth and all its components.

- **To equilibrium**: The right to maintenance or restoration of the interrelationship, interdependence, complementarity, and functionality of the components of Mother Earth in a balanced way for the continuation of their cycles and reproduction of vital processes.

- **To restoration**: The right to timely and effective restoration of living systems affected by human activities directly or indirectly.

- **To pollution-free living**: The right to the preservation of any of Mother Earth's components from contamination, as well as toxic and radioactive waste generated by human activities.

Action Three: Universal Reasonability: Breaking the Hierarchy of Care

'I believe that to meet the challenge of our times, human beings will have to develop a greater sense of universal responsibility. We must all learn to work not just for our own self, family, or nation but for the benefit of all humankind. Universal responsibility is the key to human survival. It is the best foundation for world peace, the equitable use of natural resources, and through concern for future generations, the proper care of the environment.'

TENZIN GYATSO
The 14th Dalai Lama

In the last 50 years, agro-toxins have spread and are pushing bees to extinction.

Protecting bees is an ecological duty; pushing them to extinction is an ecological crime.

The threat to bees is a threat to humanity. The choices before humanity are clear: a poison-free future to save bees, farmers, our food, and humanity in general; or continue to use poisons, threatening our common future by walking blindly to extinction through the arrogance that we can substitute bees with artificial intelligence and robots. 'Robotic bees could pollinate plants in case of insect apocalypse', ran a recent headline in *The Guardian*, reporting how Dutch scientists believe they will be able to 'create swarms of bee-like drones to pollinate plants when the real-life insects have died away'.[1]

'We see a crisis in 15 years where we don't have enough insects in the world to actually do pollination, and most of our vitamins and fruits are gone,' said Eylam Ran, CEO of Edete Precision Technologies for Agriculture. His company says its artificial pollinator can augment the labours of – and eventually replace – bees. Its system mirrors the work of the honeybee, beginning with a mechanical harvest of pollen from flowers and ending with a targeted distribution using LIDAR sensors, the same technology used in some self-driving cars.

There is no substitute for the amazing biodiversity and gifts of bees.

Every culture and every faith has seen the bees as teachers – of giving, of creating abundance, of creating the future of plants through pollination, and of contributing to our food security and welfare. The next generation of seed is transformed into the next generation of seed only through the gift of the pollinator. Navdanya's research has shown that more than 30 percent of the food we eat is produced by bees and pollinators. Nature's economy is the gift economy. In every tradition, the bee has been exemplified as a teacher in giving.

Buddhist texts note that from a multitude of living things, bees and other pollinating animals take what they need to survive without harming the beauty and vitality of their source of sustenance. For humans, to act in the manner of bees is an enactment of compassionate and conscious living. St. John Chrysostom of the Catholic Church wrote, 'The bee is more honoured than other animals, not because she labours, but because she labours for others' (12th Homily). In the Islamic tradition, the Quran's sixteenth chapter is titled 'The Bee'. This chapter is known to be the revelation of God. Bees are said to be Allah's miracles; through their behaviour, we learn to gain knowledge and reflect, to care for our community and maintain good relations. In Hinduism, there is a wonderful quote in the scripture *Srimad-Mahabhagavatam*, which reads, 'Like a honeybee gathering honey from all types of flowers, the wise men search everywhere for truth and see only good in all religions.'

Let us, together as diverse species and diverse cultures, and through poison-free organic food and farming, rejuvenate the biodiversity of our pollinators and restore their sacredness. We have the creative power to stop the sixth mass extinction and climate catastrophe without the need for these false technocratic solutions.

Here are six ways in which you can help protect the tiny bees, break the hierarchy of care and compassion, and practise universal responsibility:

1. KEEP BEE-FRIENDLY PLANTS

Honeybees need plenty of plant life to sustain their hive. Pollen from flowers is rich in protein and used to feed the larvae, while a supply of assorted nectar-rich flowers is vital to the creation of honey. A garden of (or even just a few potted) native plants that are rich in pollen and nectar is the perfect menu that will help your local bees thrive. Do some research into which flowers are good for attracting bees. Be sure to have various plants to last them throughout different seasons. If you don't have the space – or the green thumb – for a garden, an alternative is to let weeds grow into part of your yard.

2. SUPPORT YOUR LOCAL BEEKEEPERS

Your local beekeepers are doing everything they can to help their hives and promote the importance of honeybees. You can lend a hand by buying your honey locally. This will financially support your local beekeepers – and their bees. Plus, you will reinforce sustainable and environmentally friendly beekeeping practices.

Since bees are so good for the environment, buying your honey locally means you are directly supporting your local environment as well. Local beekeepers often also sell more than just honey. You can get soap, lotions, and candles made from the work of local bees.

3. GET RID OF PESTICIDES IN YOUR FOOD AND GARDEN

Cut out chemical treatments for your plants; even treatments labelled as organic could be harmful. There are various natural treatments you can try to keep pests away from your garden or yard. If you are buying new plants for your local bees, make sure the supplier didn't pre-treat them with insecticides. There are a number of petitions you can sign that demand regulation to help bees stay healthy and thriving, and promote limits on these chemicals in your state or nationwide.

4. KEEP A WATER SUPPLY

Like all living things, bees need water to stay healthy and productive. They work tirelessly to collect nectar and pollen, and if they can't find a water source, they can easily become dehydrated. You can help them out by setting up a clean, safe spot for them to rest and get a drink. This can be as pretty and extravagant as a garden pool or fountain or as simple as a small bowl filled with water. However, make sure the water is free of chemicals. It is also helpful to place stones, twigs, or other small supports in the water so bees have a place to land.

5. PROTECT SWARMS

Even though they are beneficial to our environment, bees can still alarm people, especially when we encounter them in large numbers. A honeybee swarm happens when a colony outgrows its hive and makes a mass move to a new home. Even in a swarm, bees aren't a danger when they are left alone. If you come across a swarm, leave it alone and contact local authorities. They can collect the group and relocate it to a safe, new home.

6. SPREAD AWARENESS

There are a lot of misconceptions about honeybees. When most people think of bees, they tend to think of gross or creepy insects that sting you. Get rid of these negative connotations through information, education, and awareness. This will go a long way in protecting bees and the work they do for the environment. Learn more about honeybees and share your knowledge wherever you can.

SUMMER AND RAINY SEASON GARDENS FOR BEES AND BUTTERFLIES

Yellow bhringraj (false daisy), sunflowers, mint, grass, cypress vine, holy basil, ageratum, hibiscus, oleander, kaner (*Nerium oleander*), tecoma, guava, Chinese orange flowers, gourd, beans, and brinjal flowers keep the bees busy in a garden on sunny days. The bees love to sit for a drink in the bowl of water kept for birds. The butterflies are seen more after the rains set in.

WINTER GARDEN FOR BEES AND BUTTERFLIES

Bees take to pots of mixed flowers and herbs. Mustard, saunf (aniseed), holy basil, van tulsi, black-eyed Susan, cosmos, marguerite, cornflower, chrysanthemums, candytuft, pink clover, yellow bhringraj, sunflowers, zinnia, oleander, kaner, brinjal, cockscomb, gourd, poppy, salvia, asters, penta, dahlia, and cardinal vine are a few such examples.

---

1. Maurice Maeterlinck is a Nobel Prize winner from Belgium. https://www.nobelprize.org/prizes/literature/1911/maeterlinck/biographical/

Action Four: Dissent and Democracy – Practise Ahimsa and Satyagraha

'[one] can easily conquer hate by love, untruth by truth, violence by self-suffering.'
MAHATMA GANDHI

Gandhi had started the Independence movement with the Salt Satyagraha. When the British tried to create salt monopolies, he went to the beach in Dandi, picked up the salt, and said, 'Nature has given us this for free; it was meant to sustain us. We will not allow it to become a monopoly to finance the Imperial Army.' Gandhi renewed and regenerated India's ecological civilization from the ruins of colonialism and the colonizing worldview of the imperial West. His ecological vision, which is India's civilizational vision, is even more urgently needed in our times. Gandhi always combined creative constructive action with creative resistance. For him, swaraj and satyagraha were two aspects of the movement for freedom.

Giant corporations are making huge profits at the cost of nature and people. Ordinary citizens are participating blindly as 'consumers', contributing to destabilizing the self-regulatory processes through which Gaia, the living Earth, operates. The satyagraha of our times is a 'satyagraha for life'.

As such, under the leadership of Gandhi, satyagraha turned into a global instrument of non-violent dissent against authoritarianism and a pragmatic tool of the powerless against the powerful. It refers to personal ethics and self-transformation through which an individual remains true to his or her ideals of justice and seeks to convince or convert others by working on himself or herself and taking on the burden of the sufferings of injustice. The term is often simplified, in translation, to mean 'non-violent resistance'. The resulting non-violence is not so much a practical maxim or a political strategy – although it is always political and strategic – as it is the necessary product of steadfastly staying true to one's ethical or spiritual beliefs and the ethical imperative not to hurt others.

The heart of Gandhi's ethics of interconnectedness and mutuality was to look within oneself, change oneself, and then change the world. We have given you many examples, tips, and inputs to start your journey of satyagraha. Start where you are and how you are. Be patient and gentle with yourself; as you slow down, embody the change you want to see in the world. Gandhi called his autobiography *My Experiments with Truth* for a reason. Start your own experiment with truth today.

While embarking on this journey, here are a few things to remember about satyagraha:

1. Ahimsa is a unique form of revolt that aspires to a deep self-transformation of the militant actor and a conversion of the opponents through the witnessing of self-suffering.

2. Satyagraha does not give rise to an instrumental form of non-violence but instead to an unconditional, entirely committed faith, like a spiritual belief or a moral commitment.

3. The satyagrahi can form binding moral judgments that ground her practice, but at the same time must refrain from making moral criticisms of others – even though they believe those moral judgments to be entirely right and universalizable. This severing of criticism from judgment goes hand in hand with

the satyagrahi serving as an exemplar for others, rather than criticizing them.

4. *Work on the self*: Non-violent resistance requires self-transformation. Satyagraha presupposes self-discipline, self-control, and self-purification. The willingness to bear the suffering of injustice, to take that suffering onto oneself, is at the very heart of remaining true to oneself and converting one's opponents. It is by suffering that one truly demonstrates the sincerity of one's beliefs and the stakes of justice. It is also the most powerful way to convince others to change themselves. It shows that the satyagrahi is not there to hurt but rather to impress upon others the justice of their position. Suffering is, naturally, extremely challenging. However, the fact that satyagraha is not presented as merely instrumental or strategic but rather the product of truthful belief or faith means that the actor is not constantly engaged in a reevaluation of their actions, and can remain single-mindedly focused on assuming the burden of suffering.

5. Swa-adhyan (self-study): Dedicating time to research, study, reflection, and contemplation before taking action is key for keeping democratic discourse alive. Meaningful action is not possible without this reflection, for action without reflection is mere reaction.

Here are 5 actions you can take based on Gandhian values to inspire action in your daily lives. These are also practised by the Navdanya family:

1. BEEJ SWARAJ (SEED SOVEREIGNTY)

i. Swa-adhyan (self-study). Learn about the activities of the big agri-businesses in your area and be vigilant of their actions.

ii. Boycott and encourage others to boycott genetically modified seeds while farming.

iii. Boycott products made out of genetically modified seeds – whether it be clothes, food, or any other product.

iv. Boycott the six big multinational corporations that own the world's seed, pesticide, and biotechnology industries; they have further merged – Syngenta with ChemChina, Dow Chemical (former Union Carbide with Dupont, and Bayer with Monsanto). Thus, three giants have gained control of 60 percent of the world's seeds and 70 percent of the chemicals and pesticides.

v. Launch a civil disobedience movement against any law that tries to patent indigenous seed and seed knowledge, like in the case of neem and basmati in India.

2. ANNA SWARAJ

i. Swa-adhyan (self-study). Become nutrition literate and learn about the impact of biodiversity loss, which results in the loss of diversity of nutrients in our diet that are essential to our health.

ii. Satyagraha. Demand labelling of chemicals and GMOs as a fundamental right to knowledge.

iii. Boycott genetically modified meat, dairy, and vitamin-enriched food substitutes, as their production is not energy efficient and their health ramifications unknown.

iv. Swadeshi. Eat fresh and local for your health and for local food and nutrition security by supporting nearby small-scale farmers and organic produce.

v. Sarvodaya (eat fair, eat true). Through eating fair and ensuring a decent share of what you have spent reaches the farmers, you can help end farmers' suicides.

3. JAL SWARAJ

i. Advocate for community ownership of water, ensuring access to clean, free, and easily available water for all.

ii. Don't fund the water-grabbers. Some companies and investors that buy up land around the world contribute to water scarcity and pollution. They sometimes deny local people access to water, pollute watercourses, or exhaust supplies. This can affect the ability of local communities to farm and access safe drinking water.

iii. In cities, instead of public-private partnerships (basically the privatization of water), public-public partnerships (public and government) are an alternative for the water crisis.

4. BHU SWARAJ

i. Swa-adhyan (self-study). Know your rights, especially with regard to land acquisition-related acts and forest-related acts.

ii. Stop buying products with palm oil. It is one of the biggest threats to rainforests, as its use displaces indigenous people and animals like orangutans from their land, apart from creating massive carbon emissions and contributing to forced child labour.

iii. Join a local divestment campaign. Help stop the financing of multinational fossil-fuel companies. Tell your local councillor or Member of Parliament (MP) that instead of handing our taxes over to dirty fossil-fuel companies, the money ought to be invested in renewable energy sources.

iv. Organize to demand that public money and taxes stop subsidizing fossil fuels, chemicals, and unhealthy food systems that create the burden of disease for us, and shift all public support, including policy, to health-promoting agriculture and food.

5. GYAN SWARAJ

i. Do some research. Become good at distinguishing between greenwashing and legitimate sustainability standards.

ii. Petition your government and local MPs to prohibit the patenting of all life forms – including plants, animals and parts thereof, gene sequences, and microorganisms – as well as all processes, including biological, microbiological, and non-biological, for the production of life forms and their parts.

iii. Lobby to get laws passed that protect the rights of indigenous communities and prevent any private monopolistic intellectual property claims over their traditional knowledge.

iv. Create links between schools, hospitals, healthcare centres, and local, organic, fresh, and diverse food systems.

v. Promote living knowledge – knowledge that maintains and renews living processes and contributes to the health of the planet and people. All humans have a duty to share knowledge. No person or corporation has a right to enclose or monopolize a patent, or exclusively own living knowledge as intellectual property.

vi. Collect, archive, and record indigenous knowledge and wisdom for community use and as a cherished common.

READING LIST

ADRIENNE BROWN, *Emergent Strategy.* AK Press (2017).

GLADSON DUNGDUNG, *Adivasis and Their Forest.* Adivasi Publications (2019).

VANDANA SHIVA, *Earth Democracy: Justice, Sustainability and Peace.* South End Press (2005).

GAL BECKERMAN, *Quiet Before: On the Unexpected Origin of Radical Ideas.* Crown (2022).

Recipes to Strengthen Slow Democracy

MAHUA PODA PEETHA

INGREDIENTS

250 gms mahua flowers (butter fruit)
250 gms wheat flour
Salt, to taste
Water, for batter

METHOD

- Soak the mahua flowers in water for four hours. Strain out water.
- In a medium-sized bowl, combine the wheat flour, salt, and water to prepare a batter.
- Heat the griddle/tawa over high heat. Spread half the batter on the griddle.
- Put the drained mahua flowers on the batter and cover it with the remaining batter.
- Reduce heat, and cook both sides until brown. Serve cool.

Mahua, or Madhuca longifolia, *is considered sacred among tribal people in Jharkhand because of the tree's medicinal value.*

Source: Akshat, *Navdanya*

Taking a cue from Mahatma Gandhi's life, it may be inferred that he followed a disciplined dietary regime. Brown rice was his preferred staple with dal and local vegetables. He drank goat's milk. He was a great believer in the fact that one should eat seasonal produce of raw fruits and vegetables to maximize their nutritional benefits. As a sweetener, he preferred jaggery over refined sugar.

PEANUT BUTTER AND BANANA SMOOTHIE

INGREDIENTS

3 cups dry roasted peanuts
Sea salt, to taste

Ingredients (for the smoothie)
½ cup water (or dairy-free milk of choice)
1 banana, cut into pieces
2 tbsp peanut butter paste
1 tbsp ground flax seeds
½ tsp vanilla extract

METHOD

- Make a running paste by blending roasted peanuts and salt – do not add water to it, simply keep blending it until it is runny and smooth.
- Then, in a blender, combine the water, banana, peanut butter, flax seeds, and vanilla extract. Blend until very smooth.
- You can keep adding water to help smoothen the paste and make it into the consistency you prefer.
- If you want to sweeten it further, you can add jaggery, dates, or even honey to taste.

BEES ARE OUR BEST FRIENDS – SOME HONEY-BASED HERBAL MEDICINES

Honey has been valued for its medicinal properties since ancient times. The Egyptians not only offered honey to their gods but also used it in embalming and for dressing wounds. Drawing from this long tradition, here are three remedies from a grandmother's first-aid kit worth trying:

1. For joint pain and gout: Regular consumption of nutmeg may help relieve discomfort. Add 5–6 drops of nutmeg oil to a sugar cube or combine it with one tablespoon of honey. Stir this into a glass of warm milk and sip slowly.

2. For digestive complaints: Mix half a teaspoon of fresh ginger juice with one teaspoon each of fresh lime juice and fresh mint juice, and one tablespoon of honey. This remedy is effective for dyspepsia, nausea, and vomiting caused by biliousness, indigestion from heavy non-vegetarian or fried foods, morning sickness, jaundice, and piles.

3. For fever and respiratory ailments: Combine one teaspoon of fresh ginger juice with a cup of fenugreek decoction and add honey to taste. This diaphoretic mixture promotes sweating to help reduce fever during influenza and also acts as an expectorant in bronchitis, asthma, whooping cough, and pulmonary tuberculosis.

10

SLOW CULTURE

'The great benefit of slowing down is reclaiming the time and tranquillity to make meaningful connections – with people, with culture, with work, with nature, with our own bodies and minds.'
CARL HONORÉ
In Praise of Slowness: Challenging the Cult of Speed

In this meta-universe that we inhabit today, our lifestyles have become increasingly fast, hectic, complex, and immediate. To keep up with the times, have a competitive edge and also be perceived as successful, an individual needs to earn a lot of money, show more extravagant pursuits of leisure, and be more omnipresent. This means that they must work longer hours, take higher-paying but more demanding jobs – all this so that they can spend more lavishly and consume happiness faster. If we carry on at this rate, the cult of speed can only get worse. This creates a world of competition, of enemies, and 'others'.

Samuel P. Huntington, an American political scientist and academic, once wrote, 'I cannot know who I am till I know who I hate.' The empire of greed and accumulation of the one percent (or monopoly of wealth) destroys identities, dividing societies and manufacturing enemies, to 'divide and rule' everything, everywhere. We witness this today with the wars from Iraq to Syria, Palestine, and Ukraine, as well as the growing swing towards far-right politics in most continents, leading to autocratic democracies at best and dictatorships at worst. We also see it with Facebook and Cambridge Analytica becoming the machinery of manufactured and unaccountable dissemination of hate to hijack democracy in a digital world. The currency and commodity of the empire of the one percent is hate.

Slowing down, however, creates 'oneness' based on the consciousness that we are interconnected. We are interbeings, ecologically and culturally. In these times of limitless greed and manufactured division and hate, we resist by living and thinking through our diversities as one humanity on one planet, or as the ancient Indian adage goes, *So Hum* – you are, therefore I am. There is no 'other'. Instead of identity being imposed from outside, it can flow positively from who we are, the work we do, the places and spaces we inhabit, and the relationships we nurture, and that nurture us in return. In and through this oneness, we grow gardens of diversity. We weave garlands of love.

Fast Culture vs. Slow Culture

'Fast culture is "visible, vivid, obvious and fashionable." It's like ships on the ocean that you can see and count. Slow culture "plays the country cousin, less interesting." It is everything below the surface.'

GRANT MCCRACKEN

Cultural anthropologist

There are as many definitions of culture as there are communities on this planet Earth. However, in simple terms, one can say culture is the characteristics and knowledge of a particular group of people, encompassing language, religion, cuisine, social habits, music, and arts. Etymologically, the word culture comes from the Latin word *cultura* (the tilling of land, act of preparing the earth for crops). Figuratively, it denotes 'care, culture, an honouring' from the past participle stem of *colere*, which means 'to tend, guard; to till, cultivate'.

Looking at the root of culture, we must reflect on what we are caring for, growing, or cultivating in our lives today, how that impacts our existence on the planet, and the future of the planet we inhabit. Let us look at the paradox of our times and examine aspects of fast and slow culture before we take a deep dive to understand how to build communities based on slow culture.

FAST CULTURE	SLOW CULTURE
Driven by the question How?	Driven by the question Why?
Compartmentalized existence.	Holistic living.
Hierarchy of intelligences of mind, body, and heart.	Integration of intelligences of mind, body, and heart.
Greed.	Compassion.
Self-indulgence.	Self-care.
Utilitarian linear relationships.	Healing and nurturance of a web of relationships.
Generation gap and fragmented futures.	Intergenerational wisdom and holistic futures.
Insulated communities with the fear of the other.	Celebration of diversity and open-mindedness.
Isolation and seclusion.	Interconnectedness.
Tourist in the journey of life.	Traveller in the journey of life.

So Hum: The Principle of Interdependence – You Are, Therefore I Am

'We cannot know life (jiva), without matter (ajiva). Life permeates matter, matter holds life. They are in an intricate dance of existence.'
SATISH KUMAR
You Are Therefore I am: A Declaration of Dependence

1. WANT NOT, WASTE NOT

Scarcity is caused by the depletion of resources, and this depletion is intricately linked to modern life tendencies towards wastage. Thus, wasting is tantamount to violence, as it deprives every being of resources.

2. PRACTISE HUMILITY

The consumptive and fast-paced life makes us insecure about who we are, demanding that we assert our uniqueness constantly. This insecurity breeds opinionating, showing-off our possessions, and stops us to from listening to and celebrating others, and yet we counterintuitively seek attention and validation from others. The skin-deep arrogance and forced confidence create a cognitive and spiritual dissonance, making us feel alone and empty in this hyperconnected world. Humility allows us to listen, to lead from vulnerability, and trust that we are enough by reflecting and evolving at our own pace.

3. PATHLESS TRUTH AND WEB OF LIFE

This is an invitation to reflect on values and virtues that drive our lives. Are they mere borrowed idioms and ideologies that one has acquired in life and stopped examining, thereby fencing our connections by 'othering' those who do not see the world the same way as we do? Pathless truth is encouraging us to break these silos that divide us based on the multitudes of 'isms' that we have acquired in the journey of life and encourages us to reflect through the lens of the web of life, embracing the diversity within us and around us – to use the frameworks available but not to get enslaved by them and fence ourselves in. To become human beings rather than validate our existence through the expectation of others by being 'human-doing' alone. Slow down and embrace the pathless truth and web of life.

4. HERE AND NOW

Time is not linear but sloshy, and we experience the past, present, and future in the unfolding present. So the need to be here and now and hold space is important in order to break away from the insatiable hunger of consumptive living, where as soon as we take the perfect photograph of the sunset and post it on Instagram, the beauty of the moment dies and something else needs to be consumed to feel fulfilled. Here and now, we allow ourselves to feel all emotions – joy, sadness, anger, or fear without judgement, and bathe in the vulnerability and beauty of life on Earth. It permits us space to embrace impermanence and evolution as well as the momentary stasis, allowing us to just be without our performative side overwhelming us – in short, 'just being' and feeling equanimity.

5. EMBODY INTERBEING

This refers to the recognition that all things exist in relationship with others – that our individual life and global existence are interwoven by causality that is a chain of cause and effect. That we enjoy freedom and are at the same time accountable for our actions. This is important in a world where there is an accelerated speed of entitlement to what is owed to us, with little reflection on our responsibility to others. It is an invitation to live mindfully, enjoying the simple, and what at times is deemed trivial in life, rather than being driven by daily agendas of grand tasks. That we exist and interact with all that which constitutes the planet– the trees, the birds, insects, and elements – on a daily basis, and the alchemy of this interaction impacts us every moment is the awakening into interbeing and mindful living.

5 Steps Towards Inviting and Embracing Slow Cultures

'The master in the art of living makes little distinction between his work and his play, his labour and his leisure, his mind and his body, his information and his recreation, his love and his religion. He hardly knows which is which. He simply pursues his vision of excellence at whatever he does, leaving others to decide whether he is working or playing. To him he's always doing both.'
JAMES A. MICHENER
American author

In the age of specializations and fragmented living, we have created several dualisms and walked away from an integrated life. Modern society is marred by the adage of 'work-life balance'. This distinction and compartmentalization of work and life creates abnormal expectations of life for us. We sacrifice our life at the altar of work in order to ensure a 'good life'. What we forget is that something external is defining what a 'good life' looks like for us. This materialistic understanding of a good life sends us down the rabbit hole of consumption, where our desires and life play catch-up and disappoint us completely. The question then is why do we accept this form of living where we have to strive for a 'work-life balance'? Can we imagine our existence differently and live a holistic life instead?

This idea may come as very radical to begin with, so we have broken it into small steps. The shift is one that is both philosophical and practical. Let's begin this journey together...

Action One: Slow Recreation

'I should like first to question the usefulness of the simple dichotomy of work and leisure, with work being those things which man has to do to earn his daily bread, and leisure everything he does with the time that is left over. For if we follow this way of looking at life, peculiar to our own narrow tradition, we are then faced with placing such activities as the worship of the gods, or the performance of a tragedy, in either one category or the other.'

MARGARET MEAD

Work, Leisure, and Creativity (1978)

How you relax has a deep impact not only on your mental health but also on the future of our planet. Here we will explore three fundamental shifts that we need to consider for a better future.

1. SCREEN TIME AND YOU

Yes, if one makes a list of what one does during the day while taking a break, most likely it would come down to browsing on thc internet or watching something on one of many devices. Most of us may be in the habit of watching something just before we go to bed. Many scientific studies are now sure of its impact on health. According to one such study, 'the widespread use of portable electronic devices and the normalization of screen media devices in the bedroom is accompanied by a high prevalence of insufficient sleep. Electronic devices emit an artificial blue light that can suppress the release of the body's sleep-inducing hormone, melatonin. In turn, this can interfere with the body's natural internal clock that signals when it's time to sleep and wake up. Sleep problems in early life predict a greater likelihood of later development of psychopathology in childhood and adolescence. Beyond emotional and behavioural problems, a meta-analysis of cross-sectional studies revealed that short sleep duration was reliably associated with weight gain, adiposity, and obesity risk in both children and adults.'

However, this is just the tip of the iceberg. On further analysis, one can also find a relationship between our screen time and climate change. In the article 'How smartphones are heating up the planet', Lotfi Belkhir collected data which suggests 'having conducted a meticulous and fairly exhaustive inventory of the contribution of ICT – including devices like PCs, laptops, monitors, smartphones, and tablets – and infrastructure like data centres and communication networks, we found that the relative contribution of ICT to the total global (ecological) footprint is expected to grow from about one percent in 2007 to 3.5 percent by 2020, and reaching 14 percent by 2040. That's more than half the relative contribution of the entire transportation sector worldwide.'

So what can we do to overcome this challenge? It's simple – turn off our laptops and mobiles when not in use. Many of us believe that turning off the computer is bad for the hard disk and battery, but this notion is factually incorrect. The Lawrence Berkeley National Laboratory states that modern hard drives are not affected by frequent shutdowns and that equipment may actually last longer because mechanical wear and heat stress are reduced. We also believe that our computers don't really need a lot of power if they are on but not being used. However, the truth is that during heavy usage – for example, when you open a new application – your computer draws only slightly more power. The average computer uses about 120 watts (75 watts for the screen and 45 watts for the CPU) whether you are using it or not. By simply turning off our computers and electronic devices, we save energy and reduce our dependence on fossil fuels.

2. SLOW TOURISM AND YOU

'Travel is fatal to prejudice, bigotry and narrow-mindedness.'
MARK TWAIN

Our idea of travel has changed over the years. A holiday has often become about taking photographs, due research on Tripadvisor, and figuring out everything before we even discover it. Most of our travel may also involve living in upscale properties, which have elements of local culture but the luxury and uniformity of opulence. As tourists, we are also consumers who have been trained to be greedy and gluttonous. Travelling hastily gives you a glimpse of the entirety of what a place has to offer. Hopping locations is hectic. You are on a schedule here. However, how different is it from your daily routine where you run from one task to another? Unless we slow down the travelling and make it about a journey, a discovery of both the people and things we come into contact with, remain stuck in that cycle.

Slow tourism is all about giving time to a place. Destinations have a lot to offer, and slow travel asks you to notice and engage with the people, the views and vistas, ancient local traditions, language, culture, and the way of life, along with a superb new cuisine. But only with time on your hands can you witness them all. For time is the only real currency we pay in. Only time can tell what a location can give you. There are no right or wrong answers when one is slowing down. There are only right questions to examine.

Before you take a holiday, instead of researching every location, where you want to make a pit stop at, and everything that you want to click for your Instagram profile, take some time to reflect on these questions:

- What is the purpose of your travel?

• What kind of place do you want to stay in? Does that place look at the triple bottom line of business, that is profit, people, and planet?

• What will be the ecological footprint of your travel?

• How will you impact the local people and the natural surroundings with your travel choices?

• What will you leave behind as a traveller?

Action Two: Slow Self-care and You

From ancient times, human beings have grappled with the notion of self-care, but never has it been so reactive and fast-paced as it is today. In this fast-paced life we have reduced the notion of self-care to an external expression. Self-care becomes looking after how you look, and the small joys of life get in most cases reduced to the simple act of purchasing and consuming. French philosopher Michael Foucault explains that when one embarks on the journey of self-care (what Socrates called *epimeleia heautou*), one is actually embarking on the journey of self-knowledge (or *gnõthi seauton*, the 'knowledge of the self'), which is accessed by employing various pre-established philosophical practices in all major spiritual traditions across the world. As one discovers the truth of one's being, one also discovers the mode of being and what makes one happy.

So how does one embark on this journey of self-knowledge in this fast world? The tools have existed for centuries. For instance, in ancient Greece, studying, reading, writing, note-taking, and letter writing were key to this self-knowledge. For Buddhists, it is the practice of mindfulness in which we are fully absorbed in the present moment as we watch the stray thoughts that come to

our mind without judgement while letting them go. For Advaita, it is the notion of stillness. Each philosophy suggests a different way to pause and slow down. But it is clear that without slowing down our mind and paying attention, we cannot relax or be happy, for we will be chasing a notion that we have not bothered to understand or engage with except for a momentary sense of euphoria.

For many, looking on the inside may be time-consuming, difficult, or even painful. When there is so much to take in from the outside, it almost seems counter-intuitive to think from the inside out. In our present world, there are more than enough distractions to keep us from a quiet moment of reflection, which fades fast when the phone begins to ring or perhaps causes us to feel guilty that we aren't focused on something 'more productive'. But it is important to realize that we cannot look after others and the planet if we don't know how to look after ourselves; and we cannot look after ourselves if we are fearful of facing who we are. If we do not slow down or reflect on ourselves, we will not know how to actively engage in society and politics without mere reactive responses.

Here are some simple steps we can take to slow down and know ourselves better:

1. Write regular letters to a friend(s) sharing your life's journey, things you have observed, things that bring you happiness, and those that sadden your heart.

2. Stop and smell the roses. Notice your everyday environment when you go out for a walk – observe your surroundings, your footsteps, and notice changes that take place in everyday life.

3. Find a creative expression. Write poetry, paint, do pottery, or cook and bake – just create something with your hands as often as possible.

4. Engage in deep conversations either with yourself or your friends as often as you can. Be present and listen well. Don't be in

a hurry to come to a conclusion. Digress and interconnect without the pressure of having a definite ending to the exploration.

5. Meditate on your breath everyday, even if just for 15 minutes. Notice how you are breathing. Is it shallow or deep? What thoughts come to your mind while you are trying to notice your breath? Do not judge yourself. This is not a competition, and if you ever get caught up in your thoughts, maybe you need to address them later on.

Action Three: Slow Dance to Slow Relationships

'Modern man has transformed himself into a commodity; he experiences his life energy as an investment with which he should make the highest profit, considering his position and the situation on the personality market. He is alienated from himself, from his fellow men and from nature. His main aim is profitable exchange of his skills, knowledge, and of himself, his "personality package" with others who are equally intent on a fair and profitable exchange. Life has no goal except the one to move, no principle except the one of fair exchange, no satisfaction except the one to consume.'
ERICH FROMM
The Art of Loving (1956)

A society committed to consumption and disposable objects creates relationships that are transactional, consumptive, and disposable. We cannot transform one space of life without impacting the other. Thus, how we see and interact with inanimate objects creates a psychology that impacts our relationship with animate beings as well. A society based on fulfilling short-term desires, getting rid of the broken, refusal to mend things, and deep insecurity of not having enough will create relationships that mimic the same patterns. For mending things require love and patience, while unlimited desires create dissatisfaction, and updating to the latest model reinforces that nothing is good enough for the long term. If we look at these pathologies, we often find in our reflections that relationships, whether intimate or friendships, cause us the same anxiety. Changing these patterns is hard, as our external world constantly nudges us towards the 'lacking' mentality.

BUILDING A ROAD MAP TO SLOW RELATIONSHIPS

1. PRACTISING ACTIVE LISTENING

The sense that we are not being listened to is one of the most frustrating feelings imaginable. Toddlers scream about it, teenagers move out, couples split up, and companies break down. According to the WHO, chronic exposure to noise pollution affects our blood biochemistry more than exhaust fumes by triggering the release of the stress hormone cortisol. In a push to assert ourselves or to drown out our internal monologue, we lean into adding more noise to distract us from our lives and things happening around us by being constantly plugged into earbuds, binge-watching series, or having heated and opinionated conversations. We have forgotten the art of truly listening. Improving our listening skills can help build slow relationships that are lasting and nurturing. It involves paying attention to the conversation, not interrupting and taking the time to understand what the speaker is discussing. If you really want to understand another person, you will need to tune out of yourself and your present emotions, and head towards a mode of active listening with complete focus on the other person. Any fruitful discussion or negotiation always starts with understanding the other. To be effective, active listening needs to be non-judgemental. If you start with actively listening to the other person in a non-judgemental way, you also increase the likelihood that the other person will do the same for you. This mutual listening, despite different views, can then form the basis for mutual understanding and trust, opening possibilities for reaching agreements that are beneficial for both.

2. CREATING SHARED MEANINGS

Brazilian educator Paulo Freire called the process of arriving at a shared meaning 'naming the world', in which dialogue between people created a shared sense of what the world was. The assumption that we all understand a particular word in the same way and experience it the same without discussion is one of the fundamental causes of conflict. It becomes important that we do not assume what a person's experience of situation of momentary expression is. We must take time to establish a common vocabulary and shared meaning between each other, whether it is at an interpersonal level or at the level of the collective. This process of defining the world and the place of the self within the world, according to Freire, brings about a sense of dignity. This new awareness gives us more power over the quality of our day-to-day existence. Meanings are arrived upon by reflecting on our shared values, and developing a vocabulary for our shared love, hopes, and dreams. For this, we must make time and lead from our vulnerability while trusting that this openness will be respected so that we can reflect and then act together in awareness of our frailty. Revisiting this conversation often allows us to evolve together rather than become victims of rigid and destructive patterns in our relationships. This is the process of living in reflective action rather than reactive explosions of emotions. Acceptance, compassion, and mindful transformation are the bedrocks of this process.

3. CULTIVATING STHITAPRAJNA (INNER BALANCE AND STEADFAST WISDOM)

When we define a relationship in modern times, we often refer to the notion of a spark or the 'x factor'. This flash-in-a-pan stirring of emotion that we have become attached to today becomes the yardstick of happiness. The spark of a momentary high is encouraged by consumer culture which is guided by instant gratification and immediate disposal of things and even people who do not provide it. Why mend something when you can just get a new one? This inevitably leads to a constant dissatisfaction with what one has and the fear of missing out. However, deep relationships need us to overcome this addiction to the spark and cultivate inner balance, and not be held hostage to extreme emotions and a whirlwind of action. Observing and celebrating our own inner diversity of emotions, and seeking understanding without judgement of our emotional landscape, without running away from discomfort and nurturing holistic compassion is a step towards cultivating *sthitaprajna*. It helps us overcome blaming past experiences and people while giving us hope to heal as our wisdom grows towards a balanced way of life – wherein one strives for serenity and calmness – and is not dependent on mere excitement for the sake of excitement.

Action Four: Slow Celebrations

Be conscious of your ecological impact while planning birthdays, weddings, family and friends' reunions, or any celebration. Here is a simple planning checklist to keep your celebration green and beautiful.

The Indian wedding market, estimated at approximately ₹3.78 lakh crore (as of 2022), is the world's second-largest after the US market. It is growing at an estimated 20–25 percent a year, as per a 2022 *Forbes* report. According to some studies, an average Indian spends an estimated one-fifth of their wealth accumulated in a lifetime on weddings. Weddings are largely – like most luxury purchases – about what you want to communicate to the world about yourself. More than 10 million weddings take place in India each year, but how many of them would ever be eco-friendly?

The immense cost to the environment to generate the orchestra of the grand wedding – including but not limited to the intricate invitations, wasted food, elaborate décor such as fountains, stage, services flown down from various parts of the world, heavy use of non-biodegradable products like plastic (think of the dozens of barely consumed water bottles) and fabric (synthetic and otherwise) – drowns sense and sensibility.

These are some of the steps one can take to organize a sustainable wedding or any other sustainable celebration:

1. PLANNING

The very nature of our weddings and large celebrations is geared towards waste, so one has to start from scratch to think in terms of sustainability. Proper planning of each step needs to be considered, and cost-effective sourcing and implementation will require research and choices. Hence, planning is key.

2. CELEBRATION VENUE

The first course of action is choosing the venue. For a green celebration, do not opt for upscale hotels or halls. Instead, opt for a daytime wedding in an open space like a lawn, farm, or beach, where the sun can provide the lighting, thereby reducing your carbon footprint. One can even opt for venues that have made a commitment to being ecological.

3. CATERERS AND DECORATORS

Ask your caterers to prepare food only from locally sourced ingredients. This is not only more sustainable but also fresher, tastier and, most importantly, pocket-friendly. Furthermore, insist that your caterer uses no single-use plastics at the celebration and ties up with an NGO to donate the extra food. There are many NGOs that come to collect and distribute the food at zero cost, so all you have to do is make a call! Likewise, opt for locally sourced flowers and plants for decoration. Skip all the paper and plastic.

4. INVITATIONS

For a green invitation, e-invites are the best choice. They literally cost nothing when compared to printed wedding cards, and are extremely convenient to share. At the same time, e-invites make it really easy and quick for the guests to RSVP.

5. GIFT REGISTRY

A zero-waste green wedding is just not possible without a gift registry. So create a wedding gift registry and let your guests know exactly what you'd like. Alternatively, if you have everything you need already, add a charity you support and let your guests contribute towards that cause.

6. PARTY FAVOURS

From packaging to product, support a local crafts community and always think of the ecological footprint of the product that you are giving away.

Action Five: Slow Decision-Making – Bringing Ethics and Discernment Back into our Lives

How many times have we asked ourselves: Is that the way I should treat someone else? Is that the way someone else should treat me? Due to the fact that we have the ability to be critical of our interpersonal behaviour and our contact with other beings and the physical world, we have the ability to develop codes and norms to guide that behaviour. Those moral norms and codes make our unique moral code or personal ethics. With so many challenges facing us and the planet, what can we be doing better?

Fast cultures are driven by new tools. Ethics and ecology are reduced to an irrelevant externality. The impact on society and nature is externalized. Nature and society bear the burden with ever increasing ecological degradation and degeneration of our health. Continuing on the path of disease and destruction with more invasive technologies is not intelligence – it is not *Vijnana*, scientific knowledge based on discernment between right and wrong, between what is healthy and what causes disease and ill health.

The four simple questions asked by slow cultures and democratic societies to guide our ethical compass before applying any new tool or technology in our lives are:

• Is it necessary?

• Is it better than other alternatives for our health and freedom?

• Is it safe and free of risk and hazards to our holistic health (emotional, biological, and spiritual)?

• Does it benefit nature and vulnerable communities?

For any new tool or technology, intelligence and freedom demands that we engage in the following steps to make the right choices for the health of the planet and well-being of people:

• Identify the problem.

• Take into account alternatives to address the same problem.

• Assess the efficacy of the proposed technology compared to alternatives.

• Assess the safety and unintended impacts of a new technology on biodiversity and public health on the basis of systems science and four-dimensional causality, which takes into account the complexity and interconnectedness of living systems rather than mechanistic science and Cartesian causality, which treats living systems as inert matter and reduce them to fragmented, disconnected parts. Adopt the precautionary principle to err on the side of caution in case of a lack of complete knowledge of the impact.

• Engage in democratic and intercultural dialogue to make technology choices, taking into account alternatives, and undertaking participatory, democratic assessment of necessity, ethics, ecology, efficacy, and safety.

• Ensure independent health and safety regulations and liability systems are in place before research and commercialization of new technologies which have the potential to cause harm are allowed.

Remember that in slow cultures, the well-being of all guides personal and societal change, acknowledging interdependence and compassion.

READING LIST

ERICH FROMM, *The Art of Loving*. Joanna Cotler Books (2007).

GABOR MATÉ, *The Return to Ourselves: Trauma, Healing, and the Myth of Normal*. Sounds True (2018).

BRENÉ BROWN, *Daring Greatly: How the Courage to be Vulnerable Transforms the Way We Live, Love, Parent, and Lead*. Penguin (2018).

Recipes for Slow Celebrations

PAHADI RAS

INGREDIENTS

½ cup black soy beans
¼ cup kidney beans
¼ cup black-eyed beans
¼ cup Bengal gram
¼ cup chickpeas
¼ cup black gram
½ cup horse gram
½ tsp turmeric powder
1 tsp coriander powder
1 tsp cumin powder
½ tsp red chilli powder
½ tsp garam masala
Salt, to taste
2 tbsp ghee
4–5 whole dry red chillies
A pinch of asafoetida
½ tsp cumin seeds

METHOD

- Soak all the lentils overnight, and boil them together the next day in a pot with plenty of water. Cook until the lentils are soft.
- Using a masher, mash the boiled lentils and strain the stock. Mash and strain the remaining pulp again until the water runs clear.
- In a pot over medium heat, add the stock and boil until it starts to thicken. Now add the turmeric, coriander powder, cumin powder, red chilli powder, garam masala, and season to taste with salt. Simmer the mixture until it reaches a thick gravy-like consistency. Remove from the heat.
- In a tempering ladle over medium heat, heat the ghee. Add the dry red chillies, asafoetida, and cumin seeds. Allow to crackle and pour over the prepared ras.
- Serve with rice topped with ghee.

KASHMIRI DUM ALOO

INGREDIENTS

4 green cardamom
5 cloves
2 cinnamon sticks
500 gms small potatoes
Mustard oil for frying
1½ cup yoghurt
2 tbsp gram flour/besan
1 cup water
Salt, to taste
1½ tsp red chilli powder
½ tsp dry ginger powder
¾ tsp fennel powder

METHOD

- In a mixer-grinder or spice grinder, grind the green cardamom, cloves, and cinnamon. Keep aside.
- In a pot of water over medium heat, boil the potatoes until half done. Remove from the heat and peel while they are still warm. Pierce each potato 2–3 times with a thin wooden toothpick.
- Heat the oil in a kadai over medium heat. In batches, fry the potatoes until they are golden brown.
- Drain excess oil and keep aside.
- In a medium-sized bowl, combine the yoghurt and besan with 1 cup water. Season to taste with salt. Whisk well, until smooth and lump-free.
- Reheat 4 tbsp of the same oil in another pan. Add the red chilli powder and fry for a minute. Remove from the heat and let cool. Pour this into the yoghurt, stirring continuously.
- Add the mixture to a pan and bring to a boil, again stirring continuously.
- Add the potatoes along with ginger and fennel powders and the prepared ground masala. Cook over medium heat for 5–10 minutes or until the curry thickens.

'Dum' literally means to breathe in, and 'Pukht' means to cook.

Source: Masale – Spices, *Navdanya*

PALAK CHAPATTI

INGREDIENTS

1 bunch spinach
Water to knead
1 onion, finely sliced
1 green chilli, finely chopped
4–5 curry leaves
1 cup whole-wheat flour
½ cup chickpea flour/ Bengal gram flour

METHOD

- Clean and wash the spinach and finely chop.
- Warm the water in a pan over medium heat. Add the spinach, onion, green chilli, and curry leaves to the water.
- Now add the wheat flour and chickpea flour, and prepare the dough. Knead for five minutes.
- Divide the dough into small balls. Using a rolling pin, roll the ball to make 6-inch discs.
- Heat a frying pan/tawa over medium heat. Place the chapatti on the pan and cook on both sides.

ONE OF THE MOST IMPORTANT BENEFITS OF EATING SLOW is that it gives your body time to recognize that you are full. It takes about 20 minutes from the start of a meal for the brain to send out signals of satiety. The benefits of slow eating include better digestion, better hydration, easier weight loss or maintenance, along with greater satisfaction with our meals. On the other hand, eating fast leads to poor digestion, weight gain, and lower sense of satisfaction.

Source: Kanak, *Navdanya*

INDEX

My Notes